Philosophy and Medicine

Volume 124

Founding Co-Editor

Stuart F. Spicker

Senior Editor

H. Tristram Engelhardt, Jr., Department of Philosophy, Rice University, and Baylor College of Medicine, Houston, TX, USA

Series Editor

Lisa M. Rasmussen, Department of Philosophy, University of North Carolina at Charlotte, Charlotte, NC, USA

Assistant Editor

Jeffrey P. Bishop, Gnaegi Center for Health Care Ethics, Saint Louis University, St. Louis, MO, USA

Editorial Board

George J. Agich, Department of Philosophy, Bowling Green State University, Bowling Green, OH, USA
Nicholas Capaldi, College of Business Administration, Loyola University, New Orleans, LA, USA
Edmund Erde, University of Medicine and Dentistry of New Jersey (Retired), Stratford, NJ, USA
Christopher Tollefsen, Department of Philosophy, University of South Carolina, Columbia, SC, USA
Kevin Wm. Wildes, S.J., President, Loyola University, New Orleans, LA, USA

The Philosophy and Medicine series is dedicated to publishing monographs and collections of essays that contribute importantly to scholarship in bioethics and the philosophy of medicine. The series addresses the full scope of issues in bioethics, from euthanasia to justice and solidarity in health care. The Philosophy and Medicine series places the scholarship of bioethics within studies of basic problems in the epistemology and metaphysics of medicine. The latter publications explore such issues as models of explanation in medicine, concepts of health and disease, clinical judgment, the meaning of human dignity, the definition of death, and the significance of beneficence, virtue, and consensus in health care. The series seeks to publish the best of philosophical work directed to health care and the biomedical sciences.

More information about this series at http://www.springer.com/series/6414

Joseph B. Fanning

Normative and Pragmatic Dimensions of Genetic Counseling

Negotiating Genetics and Ethics

Springer

Joseph B. Fanning
Vanderbilt University Medical Center
Nashville, Tennessee, USA

ISSN 0376-7418 ISSN 2215-0080 (electronic)
Philosophy and Medicine
ISBN 978-3-319-44928-9 ISBN 978-3-319-44929-6 (eBook)
DOI 10.1007/978-3-319-44929-6

Library of Congress Control Number: 2016955696

© Springer International Publishing Switzerland 2016
This work is subject to copyright. All rights are reserved by the Publisher, whether the whole or part of the material is concerned, specifically the rights of translation, reprinting, reuse of illustrations, recitation, broadcasting, reproduction on microfilms or in any other physical way, and transmission or information storage and retrieval, electronic adaptation, computer software, or by similar or dissimilar methodology now known or hereafter developed.
The use of general descriptive names, registered names, trademarks, service marks, etc. in this publication does not imply, even in the absence of a specific statement, that such names are exempt from the relevant protective laws and regulations and therefore free for general use.
The publisher, the authors and the editors are safe to assume that the advice and information in this book are believed to be true and accurate at the date of publication. Neither the publisher nor the authors or the editors give a warranty, express or implied, with respect to the material contained herein or for any errors or omissions that may have been made.

Printed on acid-free paper

This Springer imprint is published by Springer Nature
The registered company is Springer International Publishing AG
The registered company address is: Gewerbestrasse 11, 6330 Cham, Switzerland

Contents

1 Introduction 1
   Methodology and Terminology 3
   Debbie's Case 5
   Mapping the Project 7

2 Genetic Counseling: Models and Visions 9
   Teaching and Psychotherapeutic Models of Genetic Counseling 10
   Spiritualist Tradition 13
   A Technical Vision of Communication 19
   A Therapeutic Vision of Communication 32
   Summary 45

3 A Responsibility Model of Genetic Counseling 47
   Responsibility Model 47
   Embodiment Tradition of Communication 51
   A Pragmatic Theory of Communication 54
   Underwriting the Responsibility Model 66
   Summary 77

4 Genetic Counseling and Nondirectiveness 79
   A Brief History of Nondirectiveness 81
   Nondirectiveness and the Teaching Model 85
   Nondirectiveness and the Psychotherapeutic Model 89
   Nondirectiveness and the Responsibility Model 92
   Evaluation of Models: Debbie's Case 98
   Summary 102

5 Genetic Counseling and Spiritual Assessment 103
   Spiritual Assessment in Genetic Counseling 104
   Spiritual Assessment and Debbie's Case 124
   Summary 132

Erratum	E1
Conclusion	135
Appendix	139
Bibliography	141
Index	147

List of Table

Table 5.1　Barriers to spiritual assessment in genetic counseling 115

The original version of this book was revised. An erratum to this book can be found at
http://dx.doi.org/10.1007/978-3-319-44929-6_6

Chapter 1
Introduction

> Communication is a risky adventure without guarantees. Any kind of effort to make linkage via signs is a gamble. To the question, How can we really know we have communicated? there is no ultimate answer besides a pragmatic one that our subsequent actions seem to act in some kind of concert. All talk is an act of faith predicated on the future's ability to bring forth worlds called for. Meaning is an incomplete project, open-ended and subject to radical revision by later events. (John Durham Peters, *Speaking into the Air*)

In 2005, I observed[1] 20 prenatal genetic counseling sessions at Vanderbilt University Medical Center. With each patient's permission, I sat as a student observer in a small patient education room listening and watching the conversations that unfolded between the genetic counselors, patients, and family members. The sessions usually involved a pregnant woman who had been referred for amniocentesis[2] either because she was of advanced maternal age[3] (AMA) or because a

---

[1] This opportunity informed my research on the theoretical and ethical issues of genetic counseling and prenatal diagnosis. These sessions were not recorded nor did I take notes during the session. My observations were not intended to produce data for empirical research. Listening and watching these sessions enriched my understanding of the literature and allowed me to imagine more realistic cases.

[2] Amniocentesis is a procedure that involves inserting a long thin syringe into the woman's abdomen and drawing a sample of amniotic fluid. Before inserting the syringe, the sonographer scans to detect fetal viability, age, number, normality and position in the uterus. Knowing where the fetus is provides the optimal position for needle insertion by establishing the position of the fetus and placenta Typically done as an outpatient procedure in the 15th or 16th week but it can be done with increased risk as early as 10–14 weeks. Robert L. Nussbaum and others, *Thompson & Thompson Genetics in Medicine*, 6th/ed. (Philadelphia: Saunders, 2001).

[3] E. B. Hook, P. K. Cross, and D. M. Schreinemachers, "Chromosomal Abnormality Rates at Amniocentesis and in Live-Born Infants," *Jama* 249, no. 15 (1983): 2034–8. Hook's study and subsequent revisions by other authors indicate that the risk of chromosomal abnormalities is affected by advancing maternal age. Pregnant women who will be 35 or older at their delivery are classified as advanced maternal age by health care professionals providing prenatal care. This status entails routine referrals for a detailed ultrasound and amniocentesis. 35-years of age is significant because the risk of miscarriage from amniocentesis intersects with the risk of having a child with Down Syndrome. For revised numbers used below, see L. J. Heffner, "Advanced

screening test indicated she was in a high-risk group for having a child with a chromosomal abnormality. Initially, what attracted me to this area of research was the ethical complexity of decision making in pregnancies diagnosed with genetic abnormalities, but my observations confronted me with the equally complex phenomena of communicating about genetics. The interest in comparing and contrasting the styles of four different genetic counselors prompted the research question that guides this project: What are and what should be the dominant model(s) of communication between genetic counselors and patients?

Seymour Kessler, a leader and scholar in genetic counseling for over 30 years, describes the communicative challenges of genetic counseling this way:

> On rare occasions, the lid lifts and we are granted a fleeting glimpse into the black box of genetic counseling. What we view generally are human beings interacting and striving to understand one another. We try to overhear a few words they exchange and realize that they do not always seem to be speaking a common language. Their assumptions about things seem vastly different and there are other impediments to communication and mutual understanding. The professionals in these colloquies often seem resolved to talk about certain specific matters, numbers and statistics, for example, regardless of whatever else might be happening in the counseling interaction. Some seem to have an overriding agenda of educating the clients about the complex world of human genetics. On their part, the latter do not always seem to be certain about what they want from the professionals; their motives, wishes, thoughts and feelings seem complex and unclear, perhaps even to themselves. Communication in the session can be labored, opaque, indirect, at times incomprehensible. Clients have difficulty making themselves understood; professionals have difficulty understanding them. The result is a misdirection of efforts.[4]

Kessler's characterization invites the reader to observe with him how difficult communication and understanding are in the process of genetic counseling. Once the lid has been lifted, notice that Kessler does not begin with the image of professional and client; instead he describes two people struggling to be understood. This generalization provides a standpoint to see communication and understanding first as a human problem and second as a problem specific to professional tasks such as genetic counseling. All humans have some, if not vast, differences in their assumptions about 'things.' If it were otherwise, the need to communicate would not arise. The roles we inhabit and the spatiotemporal details of communicative acts constrain all of our efforts to be understood. Kessler pans in to show the challenges specific to genetic counseling. In his picture, professionals pursue an educational agenda that involves pre-selected content – including a genetics lesson – that lacks sensitivity to client needs. In turn, clients often lack the clarity or confidence to elicit what she or he needs from the counselor. Kessler lifts the lid not only to observe the general properties of genetic counseling but also to make evaluations about the proprieties of this practice.

Maternal AgeDOUBLEHYPHENHow Old Is Too Old?," *N Engl J Med* 351, no. 19 (2004). 1927–9.

[4] S. Kessler, "Psychological Aspects of Genetic Counseling: Xii. More on Counseling Skills," *J Genet Couns* 7, no. 3 (1998): 263–64.

This study elaborates and evaluates the proprieties of genetic counseling as they are accounted for in three models: (1) the teaching model (2) the psychotherapeutic model (3) the responsibility model. The elaboration of these involves an identification of the larger traditions, visions and theories of communication that underwrite them; the evaluation entails an assessment of each model's theses and ultimately a comparison of their adequacy in response to two important concerns in genetic counseling: the values of nondirectiveness and the recognition of differences in perspectives, specifically the response to the religious and spiritual beliefs of patients. These are discussed in reference to a case study introduced below. These theoretical efforts will ultimately support the claim that the responsibility model when underwritten by a pragmatic theory of communication provides the most adequate understanding of the proprieties of genetic counseling. Before mapping the project, a brief explication of my methodological commitments and terminological choices is needed.

Methodology and Terminology

The methodological strategy for this inquiry can be characterized as a dialectical movement *from* inferences drawn in my observations of and readings about genetic counseling and its models; *to* analysis of these inferences in relation to a normative understanding of communication; and finally a return back to specific concerns and cases in genetic counseling and their assessment in reference to analytical insights. Undertaking this strategy involves several steps at the tactical level. I identify and explicate the positions of key interlocutors who have shaped the conversation about models in the genetic counseling literature. The teaching model is articulated primarily through the work of Edward Hsia and James Sorenson; the psychotherapeutic model through the writings of Seymour Kessler and John Weil; and the responsibility model in the thought of Mary White. I elaborate and evaluate these models primarily with the insights of three theorists. H. Richard Niebuhr, John Durham Peters, and Robert Brandom. Niebuhr's notions of responsibility and sociality are key concepts in Mary White's proposal that I call the responsibility model. Peters' intellectual history provides a philosophical breadth allowing an identification and assessment of two traditions and two visions of communication that influence the models under consideration. Brandom's pragmatic theory of communication, which combines insights from hermeneutic and analytic understandings of linguistic practice, gives a detailed vocabulary that demonstrates the centrality of authority and responsibility in communication.

Four terms are used to signal moves between different levels of analysis: (1) tradition (2) vision (3) theory (4) model. If the goal of this project is the elaboration and evaluation the three genetic counseling models, then traditions, visions, and theories provide important analytical standpoints to pursue this end. The family resemblance of these terms requires stipulative definitions to show their specific usage in this study. A tradition refers to an ongoing set of general attitudes and arguments about an established phenomenon such as communication. In this study, two traditions of human communication are introduced, one is what Peters calls the "spiritualist tradition" and the other is termed the embodiment tradition.[5] From traditions we inherit the problems and solutions that constitute more specific visions of communication. A vision of human communication refers to an operational understanding of communication that can be plotted within a larger tradition. Visions can be located on a continuum somewhere between a full-fledged theory and a complete absence of reflection on communication. Visions like full-fledged theories answer important questions about the structures and functions of communication.

A vision and a theory of communication have some differences. Visions of communication are often operational in the attitudes of practitioners but seldom receive the kind of critical reflection a theoretical model undergoes in scholarly exchange. Because visions provide a manageable framework for practitioners to grasp, the practical need for scrutiny usually only comes about when there is consistent communication breakdown. Inadequate visions tend to simplify structures that are complex; globalize features that are local variations; and over- or underestimate the challenges involved in communication. Unlike a vision of communication, a fully developed theory of communication is rarely operational in the run-of-the mill concept mongering we do. Theories of communication, when diligently worked out, offer a complex set of expressive resources that allow practitioners to become self-conscious about global characteristics, e.g. perspectival difference, and local features, e.g. professional-client relationships, of discursive activity. Theoretical models can be prompted by practical circumstances such as communication failure as well as by academic aspirations to give systematic accounts of an important human phenomenon. Inadequate theories of communication tend to simplify and reify the messy retail business of discursive exchange. In this way, they can share some of the same shortcomings of visions but are expected to be defended by those who avow them.

These terminological distinctions serve the methodological goal of identifying how different levels and qualities of accounts shape the various models under consideration. A model in this inquiry refers to a schematic and normative representation of a practice such as genetic counseling. The representation consists of a distilled set of theses that articulate features of the activity such as the goals, assumptions, and tasks of genetic counseling. The models presented in this study do not seek to describe but rather guide what happens in the practice.

---

[5] John Durham Peters, *Speaking into the Air : A History of the Idea of Communication* (Chicago: University of Chicago Press, 1999), 109–36. From the spiritualist tradition, Peters moves 'Toward A More Robust Vision of Spirit' in chapter 4. He directly compares the spiritualist tradition and the robust vision of spirit as working at the same level of explanation. He also places the technical and therapeutic visions of communication within the spiritualist tradition.

Debbie's Case

The methodology of this project tests the adequacy of the theoretical terms by using them to interpret a case based on my observations. The three models are compared in relation to Debbie's case below. This case describes the contours of an actual conversation that I observed with specific details changed or added to insure anonymity. Most components of this case are unremarkable when compared with the many prenatal counseling sessions undertaken everyday in large medical centers across the nation. The circumstances of Debbie's pregnancy and her referral are common in this area of medicine. The offering of a risk assessment, a description of amniocentesis, and potential outcome scenarios are all standard parts of a prenatal genetic counseling session. These mundane qualities are strengths when comparing counseling models because the models are being applied under conditions of common practice. Two features of this case are less common but not unusual. Debbie's expression of religious concerns and the counselor's offer to leave Debbie and her spouse alone to deliberate are by no means unique or even exotic occurrences but they are not standard features of the practice.

Entering a room labeled 'Patient Education,' a 40-year old woman, Debbie, mother of two teenagers, is 16-weeks pregnant, her first time without the use of fertility treatments. She is accompanied by her spouse. She had not intended to get pregnant. The genetic counselor asks the patient to share her understanding for the referral. Debbie says that her OB/GYN referred her because of her age. The genetic counselor affirms this reason, elaborating that the patient's age puts her in a higher risk category for giving birth to a child with specific health problems. The genetic counselor does a pedigree and finds no factors that would increase the current risk assessments. She tells the patient that every pregnancy has 3–4 % background risk for birth defects and that she specifically has a risk of 1/106 for Down syndrome and 1/66 risk for any chromosomal abnormality. The patient nods her head. The genetic counselor asks Debbie whether she has any questions and Debbie indicates that she does not.

The genetic counselor asks the patient whether she knows what an amniocentesis is, reminding her that this is the test for which she has been referred. The patient indicates that she has read some information on the internet and asks whether they stick the needle through the belly button. Reassuring the patient that her belly button will not be stuck, the genetic counselor tells the patient that an amniocentesis will not be done without her informed consent and that the role of genetic counselor is to discuss what the procedure entails highlighting the risks that it carries. Having gone through the mechanics of the procedure, the genetic counselor informs the patient that the general risk of miscarriage is 1/200,[6] which is .5 % higher than the background risk for miscarriage at this stage, and the risk of serious infection is less than 1/1000.

---

[6] This risk level is the standard and lacks sensitivity to the level of experience of the physician performing the procedure.

The patient expresses her concern over putting the baby at risk and her willingness to consider abortion if the baby has Down syndrome. She says that she might terminate the pregnancy because she does not want to leave her other children the responsibility of care giving when she is gone. The genetic counselor gives her four scenarios to assist in the deliberation.

1. She can refuse the amniocentesis, avoid increased risk of miscarriage, and have a healthy baby.
2. She can refuse the amniocentesis, avoid increased risk of miscarriage, and have a special needs child.
3. She can undergo amniocentesis and miscarry a healthy baby.
4. She can undergo amniocentesis, an abnormality is found and then she must decide whether to continue the pregnancy.

The patient asks what is the probability of getting pregnant at 40 and then before the genetic counselor can answer she says that this baby is a miraculous gift. She indicates no matter what she decides that God's will would be involved adding that it would be God's will if she gave birth to a child with Down syndrome and it would be God's will if she underwent amniocentesis and a miscarriage resulted. The genetic counselor asks whether Debbie and her husband would like to be alone to discuss the options. She says yes. After 5 min, Debbie calls the genetic counselor back into the room. Debbie decides not to make a decision about the amniocentesis until she has the ultrasound results.

Many aspects of Debbie's case are generalizable and have received significant if not sufficient attention by researchers interested in the ethical, legal and social implications (ELSI)[7] of offering genetic information to patients. The category of advanced maternal age[8] (AMA) and the technological system called prenatal diagnosis[9] have been shown to affect women's attitudes about pregnancy.[10] Researchers have identified the variability in patients' understandings of risk information and genetic conditions.[11] Theologians, philosophers, and advocacy groups have

---

[7] When the Human Genome Project (HGP) formally began in 1990, the National Institute of Health and the Department of Energy dedicated a portion (3–5 %) of the HGP budget to investigate the ethical, social, and legal implications of the human genome project. This funding generated what is called E.LS.I research.

[8] R. L. Berkowitz, J. Roberts, and H. Minkoff, "Challenging the Strategy of Maternal Age-Based Prenatal Genetic Counseling," *Jama* 295, no. 12 (2006): 1446–8, R. G. Resta, "Changing Demographics of Advanced Maternal Age (Ama) and the Impact on the Predicted Incidence of Down Syndrome in the United States: Implications for Prenatal Screening and Genetic Counseling," *Am J Med Genet A* 133, no. 1 (2005): 31–36.

[9] See Ruth Schwartz Cowan's "Women's Role in the History of Amniocentesis and Chorionic Villi Sampling" in Karen H. Rothenberg and Elizabeth J. Thomson, *Women and Prenatal Testing : Facing the Challenges of Genetic Technology*, Women and Health Series (Columbus: Ohio State University Press, 1994), 35–48.

[10] Barbara Katz Rothman, *The Tentative Pregnancy : Prenatal Diagnosis and the Future of Motherhood* (New York: Viking, 1986).

[11] S. Kessler and E. K. Levine, "Psychological Aspects of Genetic Counseling. Iv. The Subjective Assessment of Probability," *Am J Med Genet* 28, no. 2 (1987): 361–70, A. Lippman-Hand and

identified many of the ethical and religious issues that arise in the connection between prenatal diagnosis and pregnancy termination.[12] One area that has received far less attention is how understandings of communication and meaning affect models of and ultimately the practice of genetic counseling. This project focuses on the relations between general accounts of communication and models of genetic counseling with the goal of establishing more adequate theoretical resources to inform better models of practice.

Mapping the Project

In Chap. 2, the teaching model and the psychotherapeutic models of genetic counseling are introduced as two dominant ways of thinking about the communication of genetic information. The claim pursued in this chapter is that these models have inherited problematic notions of communication. Appropriating the work of communication theorist, John Durham Peters, I trace this inheritance to distal philosophical stories told by Augustine and Locke and more proximate accounts whose chief narrators are Claude Shannon and Carl Rogers. I elaborate the theses of these particular understandings of communication and then identify how they operate within the respective models. Finally, I evaluate their shortcomings in an attempt to show that a different model of genetic counseling is needed.

A constructive move is made in Chap. 3. The responsibility model of genetic counseling is introduced by way of Mary White's critique of nondirective counseling and her proposal for dialogical counseling and responsible decision making. Incorporating White's insights, I offer the core elements of the responsibility model and its reliance upon a embodied, normative and pragmatic description of communication. Similar to the other two models, I locate the responsibility model within what is termed the embodiment tradition of communication. Within this tradition, I enlist Robert Brandom's pragmatic theory of communication as an effective working out of a detailed account of communication that is responsible to insights about

---

F. C. Fraser, "Genetic Counseling – the Postcounseling Period: I. Parents' Perceptions of Uncertainty," *Am J Med Genet* 4, no. 1 (1979): 51–71, A. Lippman-Hand and F. C. Fraser, "Genetic Counseling: Provision and Reception of Information," *Am J Med Genet* 3, no. 2 (1979): 113–27, J. R. Sorenson, C. M. Kavanagh, and M. Mucatel, "Client Learning of Risk and Diagnosis in Genetic Counseling," *Birth Defects Orig Artic Ser* 17, no. 1 (1981): 215–28, D. C. Wertz, J. R. Sorenson, and T. C. Heeren, "Clients' Interpretation of Risks Provided in Genetic Counseling," *Am J Hum Genet* 39, no. 2 (1986): 253–64.

[12] Erik Parens and Adrienne Asch, *Prenatal Testing and Disability Rights*, Hastings Center Studies in Ethics (Washington, D.C.: Georgetown University Press, 2000).; R. C. Baumiller, "Ethical Issues in Genetics," *Birth Defects Orig Artic Ser* 10, no. 10 (1974): 297–300, Kessler, "Psychological Aspects of Genetic Counseling: Xii. More on Counseling Skills," 263–78.; Ted Peters, *For the Love of Children : Genetic Technology and the Future of the Family*, 1st ed., Family, Religion, and Culture (Louisville, Ky.: Westminster John Knox Press, 1996).

embodiment and difference. With these expressive resources in place, I return to and develop the theses of the responsibility model of genetic counseling.

In the last two chapters, the three models are applied to two important practical concerns in genetic counseling. Nondirectiveness has been a constitutive value of genetic counseling for over 40 years but its meanings are contested. Although genetic counselors agree that patient decisions should be free of coercion, they cannot agree on a model for facilitating responsible decision making. In Chap. 4, I rehearse the history of nondirectiveness in genetic counseling and elaborate the way nondirectiveness is understood by the three models. These efforts culminate in an evaluation of how each of the models respond to Debbie's case in reference to the issue of nondirectiveness.

Chapter 5 addresses the fledgling practice of spiritual assessment within genetic counseling. After briefly introducing the relationship between spiritual assessment and the practice of medicine, I explicate and analyze the findings from two studies that explore and assess the feasibility of addressing religion and spirituality in genetic counseling. Several questions direct the analysis: (1) How should spirituality and religion be defined for the purposes of spiritual assessment? (2) Are the obstacles to undertaking spiritual assessment surmountable? (3) What are the potential benefits and harms in spiritual assessment within genetic counseling? The final move in Chap. 5 is a return to Debbie's case and the three models under consideration. I attempt to trace out their stances toward spiritual assessment and evaluate their adequacy as guides on handling spiritual and religious matters in genetic counseling.

The focus throughout this project is on models of genetic counseling and the accounts of communication they presuppose. Both of these phenomena are relatively new developments on the human stage. The ability to talk about genetics has produced great tragedies in human history and also promises to bring great benefits. The ability to talk about communication has led to insights about how linguistic practice provides ways to coordinate a shared world in the presence of real differences. My hope for what follows is that a greater awareness about the normative features of communication will allow genetic information to be coordinated with other domains of meaning that make life worth living.

Chapter 2
Genetic Counseling: Models and Visions

The professional attitudes and competencies of genetic counselors are often informed by either a teaching model or a psychotherapeutic model.[1] This bifurcation within the profession is generally accepted in the genetic counseling literature despite a diversity of modeling strategies.[2] Evidence of the pervasiveness of the two-model approach can be seen in the analytic schemes of empirical studies that distinguish educational and counseling communication styles.[3] An interest in a unified model has motivated discussion of how to combine the teaching and the psychotherapeutic models. Attempts have been made to subsume one model under another or to combine them by simple addition. In this project, I endorse an alternative model of genetic counseling; in this chapter, I claim that the teaching and

---

[1] S. Kessler, "Psychological Aspects of Genetic Counseling. Ix. Teaching and Counseling," *Journal of Genetic Counseling* 6, no. 3 (1997): 287–95.;L. J. Lewis, "Models of Genetic Counseling and Their Effects on Multicultural Genetic Counseling," *J Genet Couns* 11, no. 3 (2002): 193–212. Also see Ann C. Smith's "Patient Education" and Luba Djurdjinovic's "Psychosocial Counseling" in Diane L. Baker and others, *A Guide to Genetic Counseling* (New York: Wiley-Liss, 1998), 99–170.

[2] Ann Platt Walker, "The Practice of Genetic Counseling" in Baker and others, 1–26. In a widely used genetic counseling text book Walker identifies four models: (1) Eugenic (2) Medical/Preventive (3) Decision-Making (4) Psychotherapeutic. Whereas the concern of the present study is to analyze models that are currently operational, Walker's analysis is concerned with representing changes along a historical trajectory. More proximate is Veach, P.M., and others. "Coming Full Circle: A Reciprocal-engagement Model of Genetic Counseling Practice." *Journal of Genetic Counseling* 16, no. 6 (2007): 713–728.

[3] L. Ellington and others, "Exploring Genetic Counseling Communication Patterns: The Role of Teaching and Counseling Approaches," *J Genet Couns* 15, no. 3 (2006): 179–89.;L. Ellington and others, "Communication Analysis of Brca1 Genetic Counseling," *J Genet Couns* 14, no. 5 (2005): 377–86.;D. Roter and others, "The Genetic Counseling Video Project (Gcvp): Models of Practice," *Am J Med Genet C Semin Med Genet* 142, no. 4 (2006): 209–20. See also L. Ellington and others, "Communication in Genetic Counseling: Cognitive and Emotional Processing." *Health Communication* 26, no. 7 (2011): 667–675.

psychotherapeutic models are underwritten by problematic visions of communication that disqualify them as theoretical contenders.

Teaching and Psychotherapeutic Models of Genetic Counseling

Seymour Kessler, a proponent of the psychotherapeutic model, has summarized both approaches for the purpose of comparison. His synopsis, which will receive scrutiny below, serves as a heuristic for the rest of the project. The teaching model, according to Kessler, entails the following commitments:

1. Goal: educated counselee
2. Based on perception that clients come for information
3. The model assumes that if informed, client should be able to make their own decisions.
4. Assumptions about human behavior and psychology simplified and minimized; cognitive and rational processes are emphasized
5. Counseling task is to provide information as impartially and as balanced as possible
6. Education is an end in itself
7. Relationship with client is based on authority rather than mutuality[4]

The psychotherapeutic model, what Kessler terms the 'counseling model,' involves the following theses:

1. Goals a) understand the other person b) to bolster their inner sense of competence c) to promote a greater sense of control over their lives d) relieve psychological distress if possible e) to support and possibly raise their self-esteem f) to help them find solutions to specific problems
2. Based on perception that clients come for counseling for complex reasons
3. The model has complex assumptions about human behavior and psychology which are brought to bear in counseling
4. Counseling task complex: a) requires assessment of client's strengths and limitations, needs, values and decision trends b) requires range of counseling skills to achieve goals and c) requires individualized counseling style to fit client's needs and agendas; flexibility d) requires counselor to attend to and take care of his own inner life
5. Education is used as a means to achieve above goals
6. Relationship aims for mutuality[5]

The teaching model on Kessler's account equips the health care professional (HCP) to send an objective, unbiased message to an autonomous client who will make a rational decision once he or she possesses the right information; whereas the psychotherapeutic model conjoins HCP and patient to explore genetic information in the context of a therapeutic relationship that seeks mutual understanding as the basis for optimal decision making and adaptation. Kessler's theoretical account of the two different models has found empirical purchase in recent studies.

---

[4] Kessler, "Psychological Aspects of Genetic Counseling. Ix. Teaching and Counseling," 288.
[5] Ibid., 290.

Ellington's study of communication styles in genetic counseling sessions reinforces this two-pronged understanding of available approaches.[6] The researchers analyzed 167 genetic counseling sessions that involved explaining to the patient the circumstances and consequences of undergoing a genetic test for susceptibility to breast cancer. They identified four styles: (1) client-focused psychosocial (2) biomedical question and answer (3) counselor-driven psychosocial (4) client focused biomedical.[7] One and three were designated as consistent with the counseling model; two and four were compatible with a teaching approach. The categorizations were based on the amount of biomedical and psychosocial content discussed and the process of discussing with particular interest in who initiated what content. This study along with others indicate that there are different approaches to genetic counseling and that one way to conceptualize the differences is through the teaching and psychotherapeutic models. Although the present project focuses on the theoretical aspects of these models, empirical accounts provide valuable information about how these approaches are adopted in the performances of practitioners.

Kessler's synoptic characterization leaves little doubt that the two models, although not contradictory, entail distinct approaches to genetic counseling. The teaching model has a less ambitious agenda than the psychotherapeutic model. Under the teaching approach, the HCP primarily needs to be able to explain genetic information to different types of patients, correct misunderstandings and answer any questions the patients may have. Under the psychotherapeutic model, explaining the genetic information is only one part of a psychological equation that leads to optimal adjustment by the patient. The HCP must elicit not only the genetic history from the patient but also expressions of the patient's experience of hearing the information and other relevant collateral commitments that allow the genetic counselor to understand the patient's perspective and to intervene appropriately. Such interventions include skillful responses to the intense emotional states of some patients and to the ambivalence that some patients experience in the decision-making process. Whether these models produce the respective outcomes to which they aim is an empirical question but their differences in goals and tasks raises a normative question: How should these models be related to guide clinicians in this professional task?

One straightforward solution is to combine them. The most common route has been to subsume the teaching model under the psychotherapeutic model. Kessler's summary above attempts to incorporate the teaching model by acknowledging it as an important means to reaching larger psychological ends. The most recent definition of genetic counseling offered by the National Society of Genetic Counselors (NSGC) combines the two models and incorporates the teaching and counseling approaches seemingly without taking sides:

---

[6] Ellington and others, "Exploring Genetic Counseling Communication Patterns: The Role of Teaching and Counseling Approaches."
[7] Ibid., 183.

> Genetic counseling is the process of helping people understand and adapt to the medical, psychological and familial implications of the genetic contributions to disease. This process integrates the following:
>
> - Interpretation of family and medical histories to assess the chance of disease occurrence or recurrence.
> - Education about inheritance, testing, management, prevention, resources and research.
> - Counseling to promote informed choices and adaptation to the risk or condition.[8]

Helping people *understand* is consistent with teaching goals; helping people *adapt* follows from the psychotherapeutic model. The three kinds of implications and the three components to be integrated are an acknowledgment of both models. A question left unanswered by this definition is how to integrate all of these components if tensions exist between them. The differences between the two models entail not only kinds and quantities of goals and tasks but also the normative commitments that motivate these ends and means. Some advocates of the teaching model claim that it is inappropriate to do psychotherapy in a genetic counseling session; whereas proponents of the psychotherapeutic model claim that addressing psychological needs of the patient is a necessary aspect of genetic counseling.

In light of the NSGC definition, those who undertake genetic counseling would be expected to employ pedagogical skills such as articulating complicated information in ways accessible to a diverse client base; and to offer psychological assessments and interventions that would enhance a patient's ability to make decisions and cope with them. Kessler concludes that combining these skills into a unified approach to short-term counseling requires an "unusually gifted and flexible professional" and yet he says this is the challenge of the profession.[9] I accept Kessler's claim that utilizing this combined skill set is a challenge but question whether the details of this integration can be understood from the standpoint of either model. What both lack is an adequate account of communication that specifies the process of coordinating meanings across different perspectives.

Communication is a key term in the American Society of Human Genetics (ASHG) definition (1975) that continues to have an authoritative status in the field:

> Genetic counseling is a communication process which deals with the human problems associated with the occurrence or risk of occurrence of a genetic disorder in a family. This process usually involves an attempt by one or more appropriately trained persons to help the individual or family to (1) comprehend the medical facts including the diagnosis, the probable course of the disorder and the available management; (2) appreciate the ways heredity contributes to the disorder and the risk of recurrence in specified relatives (3) understand the alternatives for dealing with the risk of recurrence (4) choose a course of action which seems appropriate in view of their risk, their family goals and their ethical and religious standards and act in accordance with that decision and to (5) make the best possible adjustment to the disorder in an affected family member and/or to the risk of recurrence of that disorder.[10]

---

[8] R. Resta and others, "A New Definition of Genetic Counseling: National Society of Genetic Counselors' Task Force Report," *J Genet Couns* 15, no. 2 (2006): 77.

[9] Kessler, "Psychological Aspects of Genetic Counseling. Ix. Teaching and Counseling," 294.

[10] Ad Hoc Committee on Genetic Counseling, "Genetic Counseling," *Am J Hum Genet* 27, no. 2 (1975): 240–2.

The NSGC acknowledges the influence this definition has had but concludes that a more concise one is needed to circulate in the growing number of medical circumstances that require genetic counseling. The need for a shorter definition may be justified but it also increases the need to elaborate its meaning. The omission of communication from the most recent definition, by my lights, is a theoretical loss because it is the shared practice that defines the genetic counseling relationship. Some, including Kessler, welcome the jettisoning of the term 'communication process' because it seemingly refers to a mechanical transmission of information rather than a mutual relationship.[11] This transmission view of communication is impoverished and raises the question of what is an appropriate understanding of communication.

In this chapter, I propose that both models of genetic counseling are underwritten by problematic visions of communication that lack the expressive resources to be responsive to the rapidly changing contexts in which genetic counseling is undertaken.[12] These visions of communication are referred to henceforth as the *technical* and *therapeutic* visions of communication.[13] First, I rehearse what is referred to as the *spiritualist tradition* from which the technical and therapeutic visions inherit their problems. Next, I elaborate what these visions entail including the problems they inherit and their respective attempts to overcome them. I then demonstrate how the technical vision of communication underwrites the teaching model of genetic counseling and how the therapeutic vision underwrites the psychotherapeutic model.[14] The next few sections serve the purpose of relating a very specific discussion about genetic counseling to a more general conversation about communication and its challenges.

Spiritualist Tradition

The technical and therapeutic visions of communication have a more or less precise set of meanings that once set out can be traced alongside the teaching and psychotherapeutic models in genetic counseling. In telling the story of these visions and the tradition they inherit, I largely rely on the work of communication theorist, John Durham Peters. In *Speaking into the Air,* Peters ascribes a two-pronged understanding of communication in U.S. culture following World War II:

---

[11] S. Kessler, "Psychological Aspects of Genetic Counseling. Xiv. Nondirectiveness and Counseling Skills," *Genet Test* 5, no. 3 (2001): 187.

[12] I have borrowed the term 'expressive resources' from Robert Brandom. It refers to linguistic phenomena that allow us to relate explicitly to features of our world, i.e. rocks and logic, rather than remain implicit. Having expressive resources allows to talk about and judge our world in ways not possible by nondiscursive means. To recognize that expressive resources are lacking one must have access to the missing resources.

[13] Peters, *Speaking into the Air : A History of the Idea of Communication*, 63–108.

[14] In the next chapter, I introduce a responsibility model of genetic counseling and underwrite it with a pragmatic theory of communication.

In the postwar ferment about communication, then two discourses were dominant: a technical one about information theory and a therapeutic one about communication as cure and disease. Each has deep roots in American cultural history. The technicians of communication are a diverse breed, from Samuel F.B. Morse to Marshall McLuhan from Charles Horton Cooley to Al Gore, from Buckminster Fuller to Alvin Toffler but they all think the imperfections of human interchange can be redressed by improved technology or techniques. They want to mimic the angels by mechanical or electronic means…The therapeutic vision of communication in turn developed within humanist and existential psychology but both its roots and its branches spread much wider, to the nineteenth century attack on Calvinism and its replacement by a therapeutic ethos of self-realization and the self-culture pervading American bourgeois life. Both the technical and therapeutic visions claim that the obstacles and troubles in human contact can be solved, whether by better technologies or better techniques of relating and hence are also latter day heirs to the angelogical dream of mutual ensoulment.[15]

If Peters' proposal holds, then it should not be surprising that these dominant narratives about communication are found in many of the cultural practices in the U.S. especially in the biomedical sector where disease, technology, and cure are core concepts.

The two visions that Peters articulates have inherited a set of problems from what he terms the *spiritualist tradition*. The account he gives of this tradition is highly selective in its retrieval of representative texts and is almost exclusively an intellectual history. Its importance for this project is that it provides a sketch of a genealogy that traces an ongoing set of attitudes about understandings of communication. If we are to understand genetic counseling as a specific kind of communication, then Peters' account provides one interpretation of the problems and solutions that genetic counseling inherits and addresses.

Peters ascribes a movement between two basic problematics that informs the technical and therapeutic visions under consideration: "The spiritualist view of communication oscillates between the dream of shared interiorities and the hassle of imperfect media. The middle ground of pragmatic making due is rarely noted."[16]

Before cultivating pragmatic ground in the next chapter, these two problematic attitudes toward communication need elaboration. The first stance articulates the problem of human communication by comparing it to an ideal of perfectly shared interiorities. The second stance diagnoses the problem as stemming from our flawed resources for mediating interiorities.

The dream of shared interiorities is a communicative ideal that prescribes how creatures who have 'interiors' should connect. The concept of interior refers to a spatiotemporal location where attitudes, ideas, norms and preferences reside. The ideal state of connection between interiors is transparent access or complete identity with another's interior. In this state, *I* would actually be able to see the world completely from *your* perspective. The concept of interior entails an exterior. In the case of people, the exterior is bodily and hides or obstructs this interior sphere of reality. Bodies also have the property of being located in a spatiotemporal field that pre-

---

[15] Peters, *Speaking into the Air : A History of the Idea of Communication*, 28–29.
[16] Ibid., 65.

vents them from being able stand in the same spot at the same time. Our interiors or spirits are trapped and separated by this incarnate reality.

Although the intellectual history of interiority is not traced here, its expanse is hinted at by Peters' motif of angels and the role they have played in creating the dream of sharing what is inside of us.[17] Angels represent on Peters account the Christian tradition's symbol of a communicative ideal marked by the immaterial (non-mediated) contact of spirits tracing back to Augustine.[18] This ideal contrasts with the human experience of communication as a mediated, fleshly activity. The angels have no material bodies and thus do not need to incarnate their spiritual contents. But this leaves the question of whether the interior/exterior distinction applies to angels at all? Aquinas, who was more explicit about the speech of angels than Augustine, takes up this question in the *Summa Theologica*:

> External speech, made by the voice, is a necessity for us on account of the obstacle of the body. Hence it does not befit an angel; but only interior speech belongs to him, and this includes not only the interior speech by mental concept, but also its being ordered to another's knowledge by the will. So the tongue of an angel is called metaphorically the angel's power, whereby he manifests his mental concept.[19]

Aquinas notion of "interior speech" and "its being ordered to another's knowledge" can only be understood in reference to external speech between distant bodies but this order of understanding should not confused with the ordering of existence. Angels as pure spirit can logically exist and communicate prior to humanity and yet can be understood only from our standpoint as embodied spirits. What is important for this project is that Aquinas and Augustine are using a problematic ontological picture to do some important reflective work on the practice of communication. They are comparing our discursive lot to angels, and the grass is clearly greener on the angelic side.

This theme of a spiritual interior trapped inside an opaque exterior has run through many other influential narratives. A dominant reworking of this story gets expressed as the public and private domains of meaning. As Peters continues the narrative of the spiritualist tradition, the idea of interiority links Augustine to Locke who developed the notion of communication in innovative ways. Locke's individualistic account of communication combines an "Augustinian semiotic of inner and outer, a political program of individual liberty and a scientific imagination of clean processes of transmission."[20] These elements are ultimately incompatible. In trying to work out the relation between public and private meanings, Locke needed to start with the sovereignty of the individual and his ideas, but this left the public dimension of meaning under theorized.[21] His account could not square with the notion that

---

[17] The observable existence of angels is not at issue here but rather their existence within a conceptual imaginary that specifies communication.

[18] Peters, *Speaking into the Air: A History of the Idea of Communication*, 76.

[19] Thomas Aquinas and Dominicans. English Province., *Summa Theologica*, Complete English ed. (Westminster, Md.: Christian Classics, 1981). Part 1, Question 107, Article 1.

[20] Peters, *Speaking into the Air : A History of the Idea of Communication*, 88.

[21] Ibid., 87.

meanings must be understood first as social practical phenomena that get appropriated by individuals.

Nonetheless, his influence on ideas of communication can be summarized in the often-quoted statement in the *Essay Concerning Human Understanding*:

> To make words serviceable to the end of communication, it is necessary, as has been said, that they excite in the hearer exactly the same idea they stand for in the mind of the speaker. Without this, men fill one another's heads with noise and sounds; but convey not thereby their thoughts, and lay not before one another their ideas, which is the end of discourse and language.[22]

In the sentences that follow this passage, Locke does acknowledge that complex ideas cannot be replicated in the same way as simple ideas – he gives the example of moral ideas – in part because these concepts are not stable ideas within the individual. Locke clearly knew that communication rarely reached its ideal end. The acceptance of this limitation reinforced his assumptions that ideas are private and prior to their formulation in language and that language is a rough bodily instrument that fails in comparison to the sharing of pure spirits.[23]

Although Locke's project failed to reconcile its incompatible parts, Peters points out that his thinking continues to script many of the ways we understand communication. An enduring and dominant understanding of meaning, which reflects Lockean insights, is what Peregrin calls a psychologico-semiotic semantics.[24] Successful communication is one person's matching of psychic entities with signs that are then conveyed to another person who experiences the same psychic entities in recognizing the signs. The inability to communicate comes either from inadequate signs to express psychic entities or unstable psychic states that make the matching of signs difficult. The experience of not being understood or having private thoughts is easily caught up in this picture of mismatched signs. Despite the proximities of bodies and the exchange of words, spirits can remain unmatched and distant. If the ideal of communication is the complete unity of understanding as shared interiors, one is entitled to frustration with available expressive resources.

---

[22] John Locke and P. H. Nidditch, *An Essay Concerning Human Understanding*, The Clarendon Edition of the Works of John Locke (Oxford, New York: Clarendon Press, Oxford University Press, 1979), Book 3, Chapter 9, Section 6.

[23] Peters, *Speaking into the Air: A History of the Idea of Communication*, 87. Peters directs attention to Locke's comments on the communication of spirits in *An Essay Concerning Human Understanding* (2.23.36): "That, in our ideas of spirits, how much so ever advanced in perfection beyond those of bodies, even to that of infinite, we cannot yet have any idea of the manner wherein they discover their thoughts one to another: though we must necessarily conclude that separate spirits, which are beings that have perfecter knowledge and greater happiness than we, must needs have also a perfecter way of communicating their thoughts than we have, who are fain to make use of corporeal signs, and particular sounds; which are therefore of most general use, as being the best and quickest we are capable of."

[24] Jaroslav Peregrin, *Meaning and Structure : Structuralism of (Post)Analytic Philosophers*, Ashgate New Critical Thinking in Philosophy (Aldershot, Hants, England ; Burlington, VT: Ashgate, 2001), 16.

Spiritualist Tradition

This frustration leads to the second stance of the spiritualist tradition that identifies our communication problems as being caused by imperfect media. Language and the bodies that produce it are the usual suspects in discursive hassles. Many language users recognize the experience of knowing what needs to be said but not having the words to say it; or the frustration of being 'tongue tied' where the mouth and tongue as instruments do not functioning properly on occasion. In the spiritualist account, these discontents warrant either an acceptance of trapped interiors or a hopeful resolve to understand and change the problem. Locke placed confidence in the scientific knowledge of nature, ethics and communication leaving open the possibility that science and technology could overcome discursive constraints.

The problem of imperfect media gets developed in responses to the medieval scholastic question about *action in distans*.[25] How do objects/creatures at distant points on a spatiotemporal grid have an effect on one another? The concept of media, as did the English word 'communication,' emerged as scientist offered answers to this question.[26]

Scientists who addressed this question included intellectual mainstays of the sixteenth and seventeenth centuries such as Francis Bacon and Isaac Newton. Bacon offered a list of phenomena that can be transmitted such as light, heat, sound and "immateriate virtues."[27] He did not work out exactly how such transmissions took place but linked nonnormative and normative phenomena i.e. heat and virtues, in his exploration of how separated objects affect one another. Newton also wrestled to understand forces and their pathways. Peters highlights several key concepts introduced by Newton that have endured:

> Newton's description in 1687 *Principia* of universal gravitation and its operation was first and foremost an account of action at a distance. Like magnetism, light and heat, he thought gravity traveled via an "imponderable" or insensible fluid. The word Newton used for this fluid, in both his English and Latin writings, was "medium." Newton call this "universal and subtle medium" the sensorum dei (sensorium of God). He saw the cosmos as bathed in a cosmic intelligence communicating at a distance through a marvelous, intangible essence. This force or intelligence prevented us from flying off into space…Like his late nineteenth century British successors in physics, Newton took this medium not simply as a sterile physical fact but as full of spiritual suggestion. In Newton "communication" and "medium" have much of their modern senses without their modern spheres of use. One means the transmission of immaterial forces or entities at a distance and the other the mechanism or vehicle of such transmission.[28]

---

[25] Peters, *Speaking into the Air: A History of the Idea of Communication*, 75.

[26] Despite their development of notions of thought and language, Augustine and Aquinas writing in Latin did not have access to a word that plays a similar role as 'communication' does in English. Peters introduces Locke after the early scientific materialist development because he is organizing his account chronologically. I place Locke before the materialists because I am ordering my account around the two problematics of the spiritualist tradition. Etymologically, Locke inherited communication as a concept from the materialists.

[27] Peters, *Speaking into the Air : A History of the Idea of Communication*, 78.

[28] Ibid., 80.

Both Newton and Bacon were trying to understand transmissible phenomena, some observable, some not. In these efforts, they blur the line between physics of the cosmos and the metaphysics of agency. Their thoughts reinforced the link between communication and transmission and raised further questions about the material nature of communication and its mediums.

Peters characterizes Bacon's and Newton's thinking as transitional, moving away from scholasticism's highly metaphysical answers and inching closer to material explanations and technical solutions to the problem of transmission. The most important development of the link between communication and media was the telegraph. "What hath God Wrought" was the first electrical, telegraphic message sent from Washington D.C. to Baltimore by Samuel Morse in 1844. From the spiritualist standpoint, this technical feat chipped away at the spatiotemporal challenges of communication between remote bodies. Technology had overcome one kind of distance and maybe it could eventually close all the gaps between persons. An unintended consequence of the telegraph is that it reinforced the idea of disembodied spirits. As a result, a variety of cultural phenomena such the emergence of professional mediums arose that combined technological and spiritual vocabularies in a way that has continued to shape the way we think about communication. No longer did spirits seem limited by "natural" constraints.[29] The progression of technological feats in the 100 years after Morse's breakthrough reinforced the belief that humans could overcome their discursive distances either through improved mechanisms of transmission or better interpersonal techniques.

Peters' genealogy of the spiritualist tradition traces a trajectory of ideas that shapes the backdrop of specific *visions* of human communication. The dream of unmediated or transparent contact and the despair over available media may seem like remote fantasies of a bygone era; however, these dreams and dissatisfactions play a part in the ongoing political, professional, and personal conversations about responses to human differences.

The discussions about genetic counseling should be understood within this larger argument and inherit its potential benefits and harms. On the one hand, the spiritualist tradition hopes for the kind of transparency that can lead to greater understanding and responsibility. This aspiration can clearly benefit health care communication. On the other hand, it strives for a certain kind of identification with others that overrides differences requiring a kind of doubling or replication of self in the other through communication. This aim should cause concern that powerful institutions like medicine have hegemonic tendencies that undermine other vocabularies and perspectives. Keeping in mind both the benefits and the harms, the spiritualist tradi-

---

[29] Peters notes that 4 years after the telegraph the first organized version of spiritualism began in the U.S. with the rapping sounds of Kate and Margaret Fox of Hydesville, New York. These women became well known mediums that allegedly channeled the spirits of the dead. Communicating with the dead, although an ancient practice, helped to coordinate technological and spiritual vocabularies.

tion's goal of shared interiorities and its aspiration to more perfect mediums should be interpreted as motivating important visions of communication that have affected genetic counseling.

A Technical Vision of Communication

One proposed solution to our communication problems is to understand them as technical problems that require technological solutions. This solution is referred to in this project as the technical vision of communication. Mathematics is the preferred vocabulary to explain how communication works and objective information is the most promising candidate to achieve the goal of sharing interiors.

Communication theorists refer to the 1949 publication of Claude Shannon and Warren Weaver's *The Mathematical Theory of Communication* as the watershed moment not only for information theory but also for the coalescence of the technical vision of communication.[30] Fed by wartime research on cryptography and telecommunications, information theory offered a new and systematic way to think not only about technical communication systems but to think technically about communication. A message begins with an *information source* that is sent by an encoding *transmitter* through a channel with variable amounts of noise to a decoding receiver that allows the message to reach its destination. Although Shannon stated that his theory had a limited scope, his model became a powerful metaphor to describe all aspects of linguistic communication. Noam Chomsky recollects that:

> Virtually every engineer or psychologist with whom I had any contact and many professional linguists as well, took for granted that the formal models of language proposed in the mathematical theory of communication provided the appropriate framework for general linguistic theory.[31]

Why would a mathematical theory delimited to encoding, sending and decoding syntactically recognizable messages find such fertile ground in the minds of academics? The conceptual possibilities for extending the theory were easily recognizable and were developed from the start.

The process of expanding the scope of information theory begins in Weaver's introduction to Shannon's model. Weaver lists three levels of communication problems:

> Level A – How accurately can the symbols of communication be transmitted?
> Level B – How precisely do the transmitted symbols convey the desired meaning?
> Level C – How effectively does the received meaning affect the conduct in the desired way?[32]

---

[30] Claude Elwood Shannon and Warren Weaver, *The Mathematical Theory of Communication* (Urbana,: University of Illinois Press, 1949).

[31] Lily E. Kay, *Who Wrote the Book of Life? : A History of the Genetic Code*, Writing Science (Stanford, Calif.: Stanford University Press, 2000), 304.

[32] Gary P. Radford, *On the Philosophy of Communication*, Wadsworth Philosophical Topics (Belmont, Calif.: Thomson Wadsworth, 2005), 72–73.

He acknowledges that Shannon's model addressed only the problems at Level A. At the same time, he suggests that the theory could be extended to the issues at the other levels. Radford and others argue that Weaver's extension of information theory draws upon available philosophical and psychological conceptions of communication.

Weaver's first sentence of the introduction reads: "The word communication will be used here in a very broad sense to include all procedures by which one mind may affect another."[33] Radford notes that this conception of communication is consistent with the Lockean notion of communication as words exciting the same idea in the minds of speaker and audience. It should also be noted that the scholastic question about action at a distance is perpetuated in this claim. Radford cites three more passages where Weaver could not resist translating the information model to the complex practice of linguistic communication:

1) In oral speech, the information source is the brain, the transmitter is the voice mechanism producing the varying air pressure (the signal) which is transmitted through air (the channel)…
2) The receiver is a sort of inverse transmitting, changing the transmitted signal back into a message and handing this message on to the destination. When I talk to you, my brain is the information source, you're the destination; my vocal system is the transmitter and your ear and the associated eighth nerve is the receiver.
3) The semantic problems are concerned with identity or satisfactorily close approximation, in the interpretation of meaning by the receiver as compared with the intended meaning of the sender. This is a very deep and involved situation.[34]

The first two passages reinforce that Weaver is combining information theory with vestiges of a Lockean account. The third passage indirectly addresses semantic problems and associates meaning with the intentions of the sender. Radford cites a subsequent piece by Weaver that restates the problem as specifying the pathway of meaning "through some as yet unknown mind-brain process." Weaver, according to Radford, "is taking Shannon's mathematical model into the discourse of psychology, unknown mental processes, the unconscious, and the problem of discovering those information processing routines."[35] If Radford is right, then Weaver's efforts generate connections between information theory and a dominant philosophical understanding of communication as transmission; and an emerging psychological narrative about internal communication processes between the conscious and the unconscious. What Weaver demonstrates is the heritability of Shannon's model into other vocabularies. If the claim of this section is that genetic counselors are influenced by information theory and its technical vision of communication, then a story must be told about how information theory gets recombined into a genetic idiom.

Lily Kay has offered a detailed history of the influence that information theory had and continues to have on the development of the conceptual schemes in molecular biology. In a Foucauldian genealogy, she goes to painstaking lengths to show

---

[33] Ibid., 74.
[34] Ibid., 74–75.
[35] Ibid., 74.

A Technical Vision of Communication

how the powerful information and controls systems produced by a conglomeration of governmental, academic and industrial agencies in 1940s and 1950s were appropriated by molecular biologists such as Sol Spiegleman, Henry Quastler, James Watson and Francis Crick. Crick marks the borrowed status of information in an account of the central dogma of genetics:

> Once "information" has passed into protein it cannot get out again. In more detail the transfer of information from nuclei acid to nucleic acid, or from nucleic acid to protein may be possible, but transfer from protein to protein from protein to nucleic acid is impossible. Information means here the precise determination of sequence, either of bases in the nucleic acid or of the amino acid.[36]

Crick's use of information theory to understand a biological process gave birth to a host of analogies. Seminal metaphors such as the 'genetic code,' Kay points out, have become literal and hide important incongruities such as: DNA is not a code but a chain of base pairs involved in a complex process of biological specificity. The grip of the information metaphor was strengthened by the efforts of reputable linguists such as Roman Jakobson who adopted information transmission both as an adequate conception of linguistic practices and as the key to understanding biological heritability.[37] Kay's account and analysis provides insight into the conceptual resources of the academic communities that shaped genetic specialists who would eventually pioneer the task of genetic counseling.

A sketch of the technical vision's key theses will make the stance more explicit and recognizable in the teaching model of genetic counseling. Each vision and theory in this project are organized and compared by their respective answers to the following questions: (1) *What are we doing when we communicate?* (2) *What is meaning?* The answers provide criteria for recognizing how the vision motivates the respective genetic counseling model.

Theses of the Technical Vision

What are we doing when we communicate? Communication according to the technical vision is an act of transmitting information from a source to a destination. In terms of interpersonal communication, the information idiom provides a vocabulary to understand several phenomena in conversation: choosing what to say, how to say it, the speaking and hearing of an utterance, the interpretation and understanding of the utterance. As an act of transmission, conversational phenomena are roughly aligned with the processes of selecting, encoding, sending, decoding, and receiving messages. Weaver suggests that we (1) select messages from the brain (2) encode them into language (3) transmit them through an air channel by vocalizing (4) receive the message through the ear (5) decode the message from language into a

---

[36] Kay, 174.
[37] Ibid., 304–5.

form of mental state of the brain. One does not have to stretch to far to see this as a technological updating of the Lockean understanding of having mental ideas that are then matched to the linguistic signs and conveyed to a recipient who then has the same mental ideas.

What is meaning? Although Shannon's and Weaver's version of information theory claims no insight into semantics, its appropriation as a general theory of communication necessitates a theory of meaning. It is compatible with a representational theory of meaning, which is a theoretically informed account of the psychologico-semiotic theory of meaning referred to above. Meaning begins with individual words that refer to objects and properties and when combined into expressions (or sentences) represent states of affairs. This account of meaning is often characterized as atomistic in at least two respects. First, it is an explanatory strategy that begins with parts of a sentence and their corresponding referents, and puts them together to form expressions or sentences. It is also atomistic in that the meaning of one expression does not depend on other expressions but only on the sum of a sentence's parts.

Interpretation and understanding are explained in terms of decoding and receiving the meaning of an utterance or text. It should be noted here that this view runs counter to influential accounts of interpretation that can be traced to the Gadamerian insight that every text or utterance has a plurality of meanings. For the technical vision, the meaning is contained in the expression that the speaker or author forms by matching signifiers with signifieds. The circumstances and consequences of an utterance are deemphasized because the literal meaning is found in a grammatically correct form of a sentence.[38] Interpretation requires that a recipient be able to pick out (decode) expressions as sentences and have familiarity with the referents in the sentence. Education is needed in genetic counseling because patients are often not familiar with referents, i.e. chromosome.

One reason that emotions and feelings are more difficult to interpret is that what is signified remains hidden in the body of the person signifying. This might be termed a decoding problem that is compounded by the difficulty of encoding an emotional 'message'. The person who is signifying must meet the challenge of moving from a fluid awareness of a complex physiological response to an articulation of it using pre-established signifiers. As a result, the interpreter of an utterance of emotion does not have access to what the speaker is actually feeling and therefore lacks familiarity with the referent. Education is not available in the same way for this highly subjective state.

Understanding the meaning of a message is achieved in communication when the interpreter's mental state of ideas reflects the same expressions and meanings as the speaker. The same message that is sent is the same one that is received. Achieving understanding is more likely when the messages are objective in the sense of being

---

[38] John R. Searle, *Expression and Meaning : Studies in the Theory of Speech Acts* (Cambridge, Eng.; New York: Cambridge University Press, 1979), 117. Searle challenges the notion that most sentences have literal meaning independent of context or in "zero context."

observable and/or testable; understanding is unlikely when expressions signify subjective feelings and metaphysical realities.

In a strict sense, information theory is almost pure syntax.[39] It refers to a highly syntactical process of taking message forms, i.e. sentences, encoding them as numeric formulae to be decoded on the receiving end. Meaning is presupposed in the message. In terms of interpersonal communication, the technical vision proposes that the probability of effective transmission is increased when highly probable syntactic forms are sent and received. One of the challenges of genetic counseling is that complex information must be communicated in highly probable syntactic forms, i.e. forms that a patient will likely recognize. For example, if a genetic counselor carefully crafts a script with as simple sentence structure and as little jargon as possible it makes the decoding easier on the patient. Patients who can easily access and apply this information are in a better position to make reasoned decisions.

The Technical Vision and the Teaching Model of Genetic Counseling

James Sorenson's description of genetic counseling from mid 1940s to the late 1960s suggests that academic geneticists were pulled into genetic counseling because they were the only ones with knowledge to address the problems of the recurrence of genetic diseases. He characterizes their approach as "more 'scholarly' than 'consulting' professionals."[40] Using the work of medical sociologist, Eliot Friedman, Sorenson elaborates that genetic counseling was undertaken in this period with the attitude of transmitting objective information, what Friedman describes as the "ideology of technical neutrality."[41] This idea of neutrality, according to Sorenson, helped to create the principle of nondirective counseling that one finds both in the writings of early genetic counselors such as Sheldon Reed and in contemporary discussions about best practices.[42] Sorenson's account combined with Kay's history above indicates a high probability that genetic counselors at the outset understood themselves as embodying a teaching model underwritten by a

---

[39] Body language and other non-linguistic structures lack clear rules for usage and therefore tend to be neglected by this vision.

[40] J. R. Sorenson, "Genetic Counseling: Values That Have Mattered," in *Prescribing Our Future: Ethical Challenges in Genetic Counseling*, ed. D. M. bartels (New York: Aldine De Gruyter, 1993), 7.

[41] Ibid.

[42] For a review of this debate, see Sheldon Clark Reed, *Counseling in Medical Genetics* (Philadelphia: Saunders, 1955).; J. L. Benkendorf and others, "Does Indirect Speech Promote Nondirective Genetic Counseling? Results of a Sociolinguistic Investigation," *Am J Med Genet* 106, no. 3 (2001): 199–207, S. Kessler, "Psychological Aspects of Genetic Counseling: Analysis of a Transcript," *Am J Med Genet* 8, no. 2 (1981): 137–53, S. M. Suter, "Value Neutrality and Nondirectiveness: Comments On "Future Directions in Genetic Counseling," *Kennedy Inst Ethics J* 8, no. 2 (1998): 161–3.

technical vision of communication. Along with recent empirical studies, Kessler's 1997 schematic summary of the teaching model above suggests that this approach to genetic counseling persists. In rehearsing the tenets of the teaching model, Kessler relies most often on Edward Hsia's work.

Hsia's 1979 essay, "The Genetic Counselor as Information Giver" represents the clearest example of the technical visions influence on the teaching model of genetic counseling.[43] The essay appears within an influential compilation of articles that address the conceptual, social, psychological, moral and legal issues that accompany genetic counseling. In his article, which precedes Seymour Kessler's "The Genetic Counselor as Psychotherapist," Hsia claims:

> Informing a client about genetic facts and options is the *essence* of genetic counseling. The other responsibilities of a genetic counselor are essential adjuncts but these other responsibilities are not genetic counseling. In this chapter the theme will be that the focus of genetic counseling is to inform. My own attitude built on my experience as a medical geneticist and as a genetic counselor strongly favors this concept of genetic counseling as a communicative process with an educational aim. I advocate the responsibility of genetic counselors to be nondirective, nonpsychoanalytic and nonjudgmental.[44]

Hsia's interprets all the efforts of a genetic counselor as contributing to the central act of "Information Transfer from Counselor to Counselee." For this to be done effectively, the genetic counselor should gather appropriate information and know the kinds of information that patients seek.

One might expect his model to ignore counselee circumstances and psychosocial issues; instead, it incorporates these elements into the information gathering stage. In order to encode the message appropriately, a genetic counselor must assess the destination of the information by gathering several kinds of information. These include the (1) patient's prior understandings, (2) the stability of the patient's "emotional, psychiatric, socioeconomic, and family" circumstances (3) general attitudes (4) reproductive attitudes.[45] How Hsia frames this data gathering is significant. The purpose of collecting this information is to "become aware of and remain sensitive to these factors because the way in which these are addressed will determine whether the delivery of the genetic information is successful or unsuccessful."[46] What is implied in Hsia's description is that addressing the psychosocial dimensions of patients clears the channel through which the genetic information will be sent. If the patient is too unstable or too suspicious of the whole enterprise, then the genetic counselor must acknowledge that the information is not ready to be received and

---

[43] Y. Edward Hsia, "The Genetic Counselor as Information Giver," in *Genetic Counseling: Facts, Values, and Norms*, ed. Alexander Morgan Capron, Birth Defects: Original Article Series (New York: Alan R. Liss, 1979), 169–86. The 28 years that have passed since this article was published do not undermine its relevance. The importance of the article is that it presents the teaching model in sufficient detail to articulate the complexities of an approach whose prevalence is significant if not still dominant in actual practice.

[44] Ibid., 169.

[45] Ibid., 170–3.

[46] Ibid., 170.

A Technical Vision of Communication

that the patient may need to be referred to the appropriate professional for further help.

As Hsia suggests at the opening, his model has a core that involves the pure exchange of genetic information from counselor to counselee and everything else is important but peripheral. The genetic counselor should acknowledge not only the psychosocial elements of patient but also the "nongenetic" motives for seeking information:

> However earnest the genetic counselor might be about giving genetic information, when nongenetic concerns are uppermost a family will not be receptive to this information. A counselee's motives for seeking consultation can include any or all of the following[47]:

His list of nongenetic motives reinforces his strict criteria for what is and is not included in the core. Patients who seek information about causes of disease, kinds of care, natural history of a disease, and family planning are in his model pursuing nongenetic concerns. This does not mean that the genetic counselor ignores these issues. They should be addressed sufficiently to clear the channel for the genetic information. Once these concerns have been addressed then the stage has been set to transfer the information.

Hsia identifies several factors that must be considered when transferring genetic information from counselor to counselee. The first is spatiotemporal. The space must be private and free of interruptions. The dimensions and décor of the room must not be distracting. Any of these details can "hinder receptiveness of a sensitive counselee."[48] In terms of timing, he acknowledges that patient receptiveness and session length are important variables that have to be discerned in each case. The second factor considers who should be in the room. Hsia does not indicate a formula but cautions against any person who might be distracting to the patient in this case a couple, i.e. small children or a relative. The third set of factors and probably the most important is headed under, "The Manner and Content."[49]

The manner of the counselor should be "simultaneously authoritative and sympathetic."[50] Hsia elaborates on the issue of sympathy but not authority:

> a total lack of sympathy can antagonize a counselee, preventing effective transfer of information. Too sympathetic or reassuring a manner can be equally misleading because the purpose of the counseling is to convey facts and to reassure only when the reassurance is compatible with reality.[51]

Presumably, the genetic counselor must strike the same balance in undertaking authority. She must show enough authority to demonstrate competence in explaining the information and at the same time refrain from overreaching the bounds of her competencies. This restraint includes explicit acknowledgment of the limitations of genetic knowledge and accuracies of screens and tests. The correct manner

---

[47] Ibid., 173.
[48] Ibid., 175.
[49] Ibid., 176.
[50] Ibid., 177.
[51] Ibid.

needs to be matched with the right content. Having carefully planned an explanation of the genetic facts and options, Hsia cautions counselors not to give mini-genetic courses or to spend too much time on numbers because most patients are either overwhelmed or do not find these useful. Jargon should be kept to a minimum and words that evoke emotions should be avoided when possible. "The purpose of counseling is to inform without upsetting; so euphemisms are perfectly acceptable, provided they do not obscure meaning."[52] In two sentences, he notes the role of nonverbal communication claiming that it can reinforce or challenge the genetic information offered.

Although Hsia emphasizes the transfer of information *from* the counselor to the counselee, he does not ignore the need for listening and observing. He acknowledges that throughout the entire genetic counseling session the counselor must attend to nonverbal clues that express the attitudes of clients and must be "willing to be interrupted whenever the counselee wishes to ask a question."[53] The respect and interpretation of silence is also a skill that the counselor needs to hone. He notes that when the counselor talks constantly, the possibility of exchanging information is removed.

Having reviewed Hsia's model, I want to highlight features that depend on the technical vision of communication. First, his model of communication is one of transmission. This commitment is especially apparent in his consistent use of the ideas of transferring and receiving information. Most of tactics within the model serve the general strategy of removing obstacles to the patient's receptivity to genetic information. One consequence of adopting the transmission model is that a successful act of genetic counseling is defined by transferring the genetic information from the sender to the receiver with as little noise as possible. Noise in this case is the misconception of the patient, unresolved emotional issues, nongenetic issues or literally the chatter of small children. The mixed message of the Hsia's account is that all of these are important but they are not genetic counseling.

A second feature of Hsia's model that reflects the technical vision is its atomistic account of meaning. Much effort is put forth to distinguish the "essential adjuncts" of genetic counseling from genetic counseling proper, which is the explaining of genetic facts and options. Genetic information presumably has a highly circumscribed set of meanings that can be isolated from other meanings that express emotions, desires, or religious belief. His position implies that genetic counseling proper is about conveying objective information whereas its essential adjuncts include responding to subjective attitudes of the counselee. He suggests that genetic counselors should draw a bright line between facts and values and should understand that their core function is the delivery of facts not values.

This semantic account is operational in genetic counseling when the meaning of genetic information is thought of as combining references to genes/chromosomes with their properties -- or likelihood of having certain properties -- such as a mutation or translocation. Since genes and their properties are not readily observable like

---

[52] Ibid., 178.
[53] Ibid., 179.

the redness of apples, visual aids and short lessons in genetics are often used to show the meaning of genetic information. When patients comment on the genetic information in terms of feelings or personal experiences, these utterances create transmission problems. Subjective or personal messages are more difficult to encode and decode because they involve interior meanings without an objective referent. They ultimately create noise in the channel. In semantic terms, the technical vision and in turn the teaching model prioritizes extensional accounts of meaning (referents/properties to external things) over intensional ones (intentions, beliefs, feelings).

Hsia's assumptions about meaning depend on the technical vision's account of interpretation and understanding. Interpretation on Hsia's account consists of being able to receive the genetic facts and options in a form that is recognizable by the counselee. The avoidance of jargon, "flowery phrases" and emotion-laden words promote interpretation by removing unknown referents and emotional distortions from the process. A counselee has the best chance of interpreting a genetic fact when its signifiers are familiar and his or her emotions are suppressed. Successful interpretation results in understanding. This is a mental state of possessing the sentences that contained the genetic facts and options as the genetic counselor intended. Understanding is then achieved when the intended meaning of the genetic counselor is fully possessed by the counselee.

Hsia's account of genetic counseling is the most explicit adoption of the technical vision and is often referred to when discussions of the teaching model or its components arise.[54] His appropriation of the transmission metaphor enacts the technical vision of communication and allows him to articulate genetic counseling almost completely within the idiom of information theory. But is the relationship between the teaching model and the technical vision of communication a necessary one? In other words, does the teaching model of genetic counseling have to be underwritten by this particular vision? The short answer is no. In light of the rich variety of pedagogical traditions available, the teaching model of genetic counseling could be revised and underwritten by a much more robust understanding of education and communication. Such a generative effort would contribute to the value of the teaching model but would require drastic revisions that would make the model under consideration unrecognizable. I have tried to show that in the material and intellectual history there is a probable relation between the teaching model of genetic counseling and the technical vision of communication.

---

[54] Kessler, "Psychological Aspects of Genetic Counseling. Ix. Teaching and Counseling," 287–95. Kessler's work has in many ways set the terms of the discussion about genetic counseling models. His consistent reference of Hsia's work as a paradigm of the teaching model has in a sense revived Hsia's perspective 20 years later. Charles Bosk cites Hsia as a representative figures in debates about nondirectiveness in his Charles Bosk, "The Workplace Ideology of Genetic Counselors," in *Prescribing Our Future: Ethical Challenges in Genetic Counseling*, ed. D. M. Bartels, B. LeRoy, and Arthur L. Caplan (New York: Aldine de Gruyter, 1993), 27.

Evaluation

Having introduced the technical vision and demonstrated its presence in at least one influential version of the teaching model, I will now briefly propose some reasons why the teaching model of genetic counseling and a technical vision of communication are insufficient resources to comprehend the task of genetic counseling. If this model and its concomitant vision guides a genetic counselor's interaction with patients, then its content is found wanting in several areas. I will use Kessler's schematic of the teaching approach, which is based largely on Hsia's account, to assess the consequences of this model and its concomitant vision.

1. *Goal: educated counselee* – One appeal of the teaching models is its commitment to a single aim. The technical vision's account of communication reinforces this limitation on ends because the transmission model reduces the pragmatics of communication to the exchange of information. And yet Hsia's pursuit of educating a counselee faces many challenges of which he is fully aware. His explanatory strategy is to organize genetic counseling as having a core function or essence and a set of essential adjuncts. The former entails the explaining of facts and options; the latter involves everything else.

   The strength of Hsia's conceptualization is that it does articulate the connection between addressing essential adjuncts and offering genetic information. Patients are sometimes not ready to process information because of personal circumstances related to emotional, family or economic circumstances. Moreover, many counselees do come with misconceptions that need to be addressed. The problem with acknowledging this plurality of issues is that it becomes more difficult to draw lines between content. Hsia's development of the connection between receiving information and the psychosocial state of patient is in tension with his initial claims about the essence of genetic counseling. If a genetic counselor decides that a patient is not ready to receive genetic information referring her instead to a social worker, then has any genetic counseling taken place? In Hsia' strict sense, it has not because genetic information has not been transferred, no counselee has been educated. If such an interaction is not genetic counseling, then how should one understand the genetic counselor's assessment of a counselee's emotional readiness to be educated? Presumably, Hsia would characterize it as a rudimentary diagnostic skill that allows a genetic counselor to recognize whether a counselee is "in a state" to be receptive. The single mindedness of this model misses the importance of the variety of interactions that occur within a genetic counseling session and also the plurality of benefits that are in play.[55]

2. *Based on perception that clients come for information* – The teaching model might justify its restrictive view of genetic counseling by referring to its percep-

---

[55] For a study that identifies the large number of benefits at stake, see, B. A. Bernhardt, B. B. Biesecker, and C. L. Mastromarino, "Goals, Benefits, and Outcomes of Genetic Counseling: Client and Genetic Counselor Assessment," *Am J Med Genet* 94, no. 3 (2000): 189–97.

tion that patients come to receive genetic information rather than psychotherapy or moral consultation. Kessler's synopsis of the model is guilty of being too simplistic here because Hsia acknowledges that counselees' attitudes are often motivated by nongenetic concerns. A better description of the model's stance is that the client *should* come to receive genetic information. If he or she does not, then the genetic counselor's task is to address other concerns in order to clear the way for the transfer of genetic information. The problem with attributing *the* reason a counselee should participate in the session is that the other reasons may actually be vital to a counselee's understanding of the genetic information in a holistic way. Hsia's account presupposes an atomistic semantics where meanings can be contained in their separate spheres, whereas the conversational moves in an actual session do not function within such circumscribed spheres. This feature of conversation challenges the notion that districts of meaning can be easily defined. For example, in Debbie's case, the genetic counseling session refers to chromosomes, responsibility, and God's will. These seemingly disparate meanings are all tied to the interpretation of genetic information. I will say more on this below.

3. *The model assumes that if informed, clients should be able to make their own decisions* – Although the issues around decision making are taken up more fully in Chap. 4, several problems can be suggested at this stage. Hsia has a highly restrictive view of the counselor's role in a counselee's decision making process. His hands-off approach is motivated in part by his definition of a rational decision. Once the information has been transferred then the patient must make a decision with respect to his or her beliefs. A rational decision is one that is made after careful deliberation of the genetic information by the patient. The attribution of rationality is not based on outcome. The strength of Hsia's view is that 'rational' is not restricted to logical inferences that follow from an idealized scientific perspective. Rational decisions are marked by the patient's reasoning and cannot be made by the genetic counselor even when the patient wants or expects a recommendation from the health care professional. The weakness of Hsia's view is that it advocates the presumption that a counselee is fully transparent to herself and is fully competent in making inferences about genetic information using personal information. Hsia's model presumes that interiorities can be shared only through a thin channel of objectivity and that the subjective contents of decision making are better left to their owners.

4. *Assumptions about human behavior and psychology simplified and minimized; cognitive and rational processes are emphasized* – Hsia's account does simplify assumptions about human psychology or behavior. This effort to minimize such assumptions is an acknowledgment of the limitations of a genetic counselor's competencies in understanding and responding to the psychological and social problems of a patient. At the same time, these concerns, Hsia notes, cannot be avoided. Hsia's account glosses over what competencies are required to be sensitive to these issues in a way that serves patient education.

The teaching model does emphasize cognitive and rational processes but the problem is that this emphasis is simplistic. The presuppositions about the interpretive processes and the achievement of understanding rely on the technical vision's image of transfer and receipt with the result of achieving identical mental state in reference to the genetic information. The complexities and challenges of linguistic practices are reduced to transmission problems that are attributed to generic differences between perspectives. That vocabularies, i.e. religious and genetic, themselves are sometimes difficult to adjudicate is never considered.

5. *Counseling task is to provide information as impartially and as balanced as possible* – Most of the teaching model's commitments hinge on the idea that under the right conditions a genetic counselor can transfer objective information to the patient in a way that the patient can receive it. A criticism of this assertion is that the genetic counselor decides what information should be given and that this selection is based on perceptions about what information is valuable. Hsia acknowledges that a completely unbiased counselor is a "myth":

> Whenever a counselor presents information, however sincerely he or she may strive for objectivity, the tone of voice, choice of words and nonverbal will all add subjective color to the objective facts.[56]

This admission maintains a distinction between objective facts and their delivery. What Hsia's model misses is the differences between the circumstances that produced the objective fact and the circumstances that entitle a genetic counselor to offer it to the counselee. What makes a fact objective is the repeatable circumstances that produced the proposition. The circumstances that entitle genetic information to be offered to patients are not the same as the circumstances of its production as a fact. For example, there has been significant debate about when an amniocentesis should be recommended to a pregnant woman. The criterion currently in use is: When the probability of miscarriage from amniocentesis is greater than or equal to the probability of having a child with a genetic abnormality, then the pregnant woman should offerred amniocentesis. These two outcomes can be compared because they can be interpreted as harms. The circumstance in which the information is offered is clearly not the same as the circumstance that produced the probability nor are the potential harms commensurate. The teaching model as expressed by Hsia underestimates the values needed to use objective facts in circumstances outside the ones that produced them.

6. *Education is an end in itself* – This commitment has similarities to the goal of an educated client but is a much stronger claim, one that is highly compatible with the technical vision of communication. One consequence is that the criterion for success of genetic counseling becomes transferring the genetic information in an accessible way and answering all of the patient's questions. Confirmed success could be established through an objective instrument to measure the counselee's

---

[56] Hsia, 184.

understanding.[57] A second consequence is that other concerns can be dealt with as issues subservient to the educational task similar to the way classroom teachers sometimes view disciplinary issues.

The problem with this restrictive model of communication can be seen in following circumstance: The genetic counselor has succeeded on the above terms but is left with an ambivalent patient who is clearly distraught by what they have learned. To define success in purely educational terms raises the question of the genetic counselor's status as a health care professional and the benefits this role should attempt to confer. Is the genetic counselor a technician of information or a professional charged with the responsibility of caring for a person and their health? Given that genetic counseling is the task of HCPs, the claim that education is an aim in itself is incompatible with the other values and responsibilities traditionally attributed to clinicians.

7. *Relationship with client is based on authority rather than mutuality* – This ascription by Kessler is in part true in reference to Hsia's account. The claim that genetic counseling should be purely educational entails a certain kind of authoritative structure of communication. A knowing counselor imparts knowledge to an unknowing counselee. Here again Hsia's implicit understanding of semantics supports his view that the genetic counselor, if nothing else, has control over the meaning of genetic information. Hsia would acknowledge that the counselee has control over decision making; his model appears less equipped to acknowledge that the patient's perspective plays a role in what the genetic information means. I can only assert that this is problematic here and will develop this semantic point in the next chapter.

Another problematic feature of Hsia's account is that the dispositions of authority and sympathy are deployed exclusively in the service of educational benefits. He states that "a total lack of sympathy can antagonize a counselee, preventing effective transfer of information."[58] Elsewhere he asserts the need for the counselee to trust the counselor but this trust does not entail the belief that the counselor cares about the counselee in terms of health. Instead it is a trust in the counselor's knowledge of the information and transparency about the limits of his or her knowledge. He reduces the motivation for showing sympathy to the optimization of information transfer. Such a reduction makes his model susceptible to charges of being overly cognitive and even more serious of being manipulative. More will be demonstrated about the shortcomings of the teaching model in Chaps. 4 and 5.

---

[57] One line of research within the genetic counseling tradition is outcome based. For a review of these issues, see A. Clarke, E. Parsons, and A. Williams, "Outcomes and Process in Genetic Counselling," *Clin Genet* 50, no. 6 (1996): 462–9.

[58] Hsia, 177.

A Therapeutic Vision of Communication

The technical vision of communication addresses the problem of sharing interiors by specifying and creating conditions of high probability for accurately transmitting information. It looks to objective information and its accessible referents as the most likely conceptual content to be transferred effectively. The suggestion that Weaver's extension of the technical vision was moving toward psychology points to the availability of psychological discourses about communication. In this section, a psychological account of communication is offered in terms of Carl Roger's humanistic psychology. After rehearsing its components, the therapeutic vision's impact on genetic counseling is demonstrated.

The therapeutic vision has roots prior to World War II but came to fruition after the war in part through the writings of psychotherapists like Carl Rogers. Because Rogers has had a traceable influence on genetic counseling, I will explicate the therapeutic vision in terms of the major commitments that he put forward. Peters reinforces the choice of Rogers as spokesperson for this distinctive view of communication:

> Carl R. Rogers, the leader of person centered, humanistic psychology in the postwar era is perhaps the best example of a therapeutic theorist of communication. As he put it in a talk in 1951, "The whole task of psychotherapy is the task of dealing with a failure in communication." Communication breakdown for him was the fate of the neurotic, whose communication both with himself and with others was in some way damaged – blockage of communication occurring between the unconscious and the ego, for instance. "The task of psychotherapy is to help the person achieve, through a special relationship with a therapist good communication within himself." Good communication with others would follow. As Rogers summarized, 'We may say then that psychotherapy is good communication, between and within men. We may also turn that statement around and it will still be true. Good communication, free communication within or between men is always therapeutic.'[59]

This characterization of Rogers captures both the psychological bent to the vision and also its universal aspiration. A key psychological concept in his view of communication is congruence. Through a brief elaboration of this notion, we can begin to trace the trajectory of his thought.

In *On Becoming a Person,* Rogers defines congruence as an "an accurate matching of experiencing and awareness. It may still be extended further to cover a matching of awareness, experience, and communication."[60] An infant who is moved by an empty stomach to cry is according to Rogers in a unified state of congruence: the experience and awareness of hunger and the communicative response are integrated. His counter example involves a man who is exhibiting anger in a conversation and who denies this anger when confronted. This man is in a state of incongruence, not aware of his anger even though the anger is explicit in his tone of voice. Congruence is the compatibility between operational attitudes about internal

---

[59] Peters, *Speaking into the Air: A History of the Idea of Communication*, 26.

[60] Carl R. Rogers, *On Becoming a Person; a Therapist's View of Psychotherapy* (Boston,: Houghton Mifflin, 1961), 339.

and external stimuli and one's attitudes about those attitudes. The therapeutic vision of communication, in Rogers' view sets congruence as its end and client-centered therapy as its means.

Client-centered therapy, a name Rogers would later change to person-centered therapy, is the set of practical attitudes and implementations[61] that helps others achieve congruence. The root hypothesis motivating these attitudes is: Every person has the capacity to be autonomous. Autonomy for Rogers is the ability to choose one's own goals, to decide how to reach those goals and to take responsibility for both the goals and the decisions. A major obstacle to this autonomy is the fear of experiencing who one truly is. This fear prevents a person from learning her true feelings, desires, and interests that ultimately motivate authentic goals. When this fear subsides, a channel of communication is opened up:

> Often I sense that the client is trying to listen to himself, is trying to hear the meanings and messages which are being communicated by his own physiological reactions. No longer is he fearful of what he may find. He comes to realize that his own inner reactions and experiences, the messages of his senses and his viscera are friendly. He comes to want to be close to his inner sources of information rather than closing them off.[62]

Being open to communication with one's true self is essential to achieving the congruence required to authentically undertake one's autonomy. Attributing autonomy and the possibility for congruence to the client is on Roger's view the starting point for therapeutic communication.

This attribution takes specific form in the therapeutic process through empathic identification. Rogers characterizes the stance this way:

> The counselor's function is to assume in so far as he is able the internal frame of reference of the client to perceive the world as the client sees it, to perceive the client himself as he is seen by himself, to lay aside all perceptions from the external frame of reference while doing so and to communicate something of this empathic understanding to the client.[63]

The effort to understand the client requires an acceptance of the interior that comes into view through the process. If empathic identification functions like a mirror that helps the client see what needs to be rearranged, then evaluations or judgments by the counselor distort what the client sees. Counselors who employ the Rogerian approach in as much as possible strive to share the client's interior in a way that allows not only the therapist but also the client to have otherwise unavailable access. The client gradually discards prior filters of experience, moves to an immediacy of experience and then gradually replaces old attitudes with provisional ones that are more sensitive to actual experience. The result is that "internal communication is clear with feelings and symbols well matched and fresh terms for new feelings."[64]

---

[61] Rogers avoids using the terms 'techniques' or 'methods' because of the concern that when detached from the root hypothesis clients quickly recognize and resist them.

[62] Rogers, 174.

[63] Carl R. Rogers, *Client-Centered Therapy, Its Current Practice, Implications, and Theory* (Boston,: Houghton Mifflin, 1951), 29.

[64] Rogers, *On Becoming a Person; a Therapist's View of Psychotherapy*, 154.

The dream of shared interiors, it seems, has partially been obtained through the therapeutic process. The therapeutic vision of communication offers access to interiors that can be seen as they truly are without the distortion of prior dogma and external biases. Before explicating the Rogerian program in terms of the two questions, an account about how this vision entered into the field of genetic counseling needs to be sketched.

In 1969, Sarah Lawrence College accepted ten students into the first academic institution to offer a program specifically designed to train Masters-level genetic counselors. Melissa Richter, the professor who took the steps to open the program had training in biology and psychology and believed that clients receiving genetic information needed professionals who could attend to their psychological needs. In 1976 Sarah Lawrence offered a course titled 'Client-Centered Counseling.' Joan H. Marks, who directed the Sarah Lawrence program for 26 years until 1998 and was trained in psychiatric social work, described the course as "based on Carl Roger's concepts of nondirective counseling, employing empathic responses against the background of unconditional positive regard"[65]

As recently as 2008, genetic counseling students at Sarah Lawrence took a course titled "The Empathic Attitude" taught by Marvin Frankel. The course description is representative of the therapeutic vision and reflects a Rogerian influence:

> This course provides a theoretical and practical understanding of client-centered counseling. Students participate in tape-recorded interviews with role-playing subjects, which provide the basis for subsequent classroom discussion. Rogerian techniques are applied and integrated into clinical genetic counseling cases. Special emphasis is placed on understanding the emotional content of language in all phases of the genetic counseling process; eliciting a client's psychological needs; and the choice of vocabulary in explaining complex genetic phenomena.[66]

Combined with a substantial regimen of genetic courses, students at Sarah Lawrence learned what it means to communicate empathically with clients about their experience of receiving genetic information. This course description suggests the continuation of the trajectory of the therapeutic vision of communication whose entrance into the field of genetic counseling is clearly documented in the program at Sarah Lawrence and its influence in the field. Sarah Lawrence's program continues to influence the field in the total number of graduates working as genetic counselors and leading other genetic counseling programs.[67]

---

[65] Joan Marks, "The Training of Genetic Counselors: Origins of a Psychosocial Model," in *Prescribing Our Future: Ethical Challenges in Genetic Counseling*, ed. D. M. Bartels, B. LeRoy, and Arthur L. Caplan (New York: Aldine de Gruyter, 1993), 20.

[66] Sarah Lawrence College, *Human Genetics 2007–2008 Courses* [website] (Sarah Lawrence College, 2008, accessed January 11 2008); available from http://www.slc.edu/human-genetics/Courses.php.

[67] For a compelling account of Sarah Lawrence's influence, see Arno Motulsky, "2003 Ashg Award for Excellence in Human Genetics Education: Introductory Remarks for Joan Marks," *Am J Hum Genet* 74 (2004). See also Stern, A. *Telling Genes : The Story of Genetic Counseling in America*. Baltimore: Johns Hopkins University Press, 2012.

A Therapeutic Vision of Communication

Having established the link between the therapeutic vision of communication and genetic counseling, an elaboration of the therapeutic vision of interpersonal communication in terms of our two questions provides some standard criteria to recognize it within the genetic counseling literature.

Theses of the Therapeutic Vision

What are we doing when we communicate? Interpersonal communication is the sharing of each person's individual experience through the matching of appropriate symbols. Evaluative and judgmental attitudes distort the process of communication because they obstruct our ability to empathically identify with the other person. Interpersonal communication presupposes intrapersonal communication that occurs between a person's experience and her awareness of the experience. Experience is an "organismic event" and awareness is a construal or symbolization of this event. The therapeutic vision claims that communication between persons is greatly enhanced when congruence, or accurate symbolization of experience, is present in each of the persons.

What is meaning? Like the technical vision, the therapeutic vision offers a representational theory of meaning. Meanings are built by matching symbols to experiences. Symbols with varying degrees of accuracy are attached to and filter experiences of perceived reality.[68] Experience is the interaction of subconscious input from a phenomenal field and conscious awareness that allows for symbolization. Each person's symbol set functions like a map that helps them navigate the phenomenal field.

Unlike the technical vision, the therapeutic vision focuses on the internal referents of experience as it gets expressed in feelings and beliefs and the accuracy of the concepts attached to them. Communication guided by therapeutic attitudes permits "faulty and generalized symbols" to be "replaced by more adequate and accurate and differentiated symbols."[69] The latter symbols emerge through a process of testing experience. The accuracy of meaning comes from testing the symbolization of private experience; however it does not follow that intensional forms of meaning are preferred.

When the therapeutic vision explicitly adopts semantic terms, it parallels the technical vision's preference for extensional over intensional forms of meaning. On this view, intensional qualities of meaning are marked by the "tendency to see experience in absolute and unconditional terms, to over generalize, to be dominated by concept or belief, to fail to anchor his reactions in space and time, to confuse fact

---

[68] Rogers, *Client-Centered Therapy, Its Current Practice, Implications, and Theory*, 144–5.
[69] Ibid.

and evaluation, to rely on ideas rather than upon reality-testing."[70] Unobstructed by dogmatic filters that block communication between experience and awareness, accurate symbolization has an extensional quality that is tested against direct experience of the phenomenal field rather than against a highly abstracted map.[71]

The link between meaning and interpretation has two important dimensions. First, interpretation is a form of matching symbols and experience. The therapeutic vision claims that fear of the true self prevents persons from matching appropriate symbols to immediate experience. The fear is rooted in abandoning dogmatic pictures of the self and replacing them with provisional but more responsive conceptions of individuated experience. This leaves open the question of where the new conceptions come from. True interpretation of one another starts on Roger's view at home in the accurate matching of symbols with internal experience, what he calls internal communication. Thus, an authentic interpretation of others requires (1) that the speaker's symbolic utterance is appropriately matched with his or her experience (2) that the listener's receipt of the utterance entails an empathic openness to content. When the receiver hears or sees the sentence, then he or she attempts to match it with the imagined interior state of the speaker or author. This interpretive scheme culminates in the ideal state of mutual understanding where I understand the world from your perspective and you understand it from mine.

Linguistic structures, i.e. syntactically recognizable utterances, can express or conceal true meanings. When symbols are properly matched to an individual's experience of the phenomenal field, then syntactic structures are a valuable tool for sharing experiences. In circumstances where an incongruence can be detected, nonverbal communication usually indicates the more reliable experience of the person. Although nonverbal communication can be less determinate than verbal expression, the fullness of understanding another person is not complete without interpreting their bodily reactions in relevant circumstances.

The Therapeutic Vision and Psychotherapeutic Model of Genetic Counseling

The psychotherapeutic model of genetic counseling explicitly avows features of the therapeutic vision of communication as can be seen in the work John Weil and Seymour Kessler. Most advocates of the psychotherapeutic model avow three components that are directly borrowed from a Rogerian approach: unconditional positive regard, empathy, and genuineness. John Weil, a leading proponent of this approach to genetic counseling, refers to a passage by Bohart to elaborate these "three critical elements":

---

[70] Ibid., 144.
[71] Ibid.

A Therapeutic Vision of Communication

Unconditional positive regard involves respecting and accepting the counselee as a complete individual including his or her strengths, weaknesses and full range of feelings and behaviors. While it is unconditional with respect to the individual, it need not be so with respect to specific behaviors or aspects of personality. Thus, the counselor acknowledges to herself that the counselee has some less positive aspects and she addresses them in an appropriate manner. This includes setting limits on unacceptable or threatening behavior

Empathy involves an understanding insofar as possible of the counselee's lived reality. This includes his or her past and present experiences, emotions and perceptions of the world and the role these play in shaping behavior.

Genuineness involves the counselor's openness to her own emotional experiences in the interaction with the counselee and a modulated but honest expression of this in her interaction with the counselee.[72]

When one of these is missing, according to Weil, the message being sent to the counselee is that emotional issues will not be addressed. Even more serious, is the tacit message that emotional issues are "too scary to address or too abnormal to be tractable or that the bearer is too needy, pathological or unacceptable to be helped."[73] When these attitudes are embodied by the counselor, the circumstances are established for helping the patient to address, understand, and adapt to the emotional states that result from receiving genetic information.

Fulfilling these attitudes depends on informed observation of the counselee's verbal and nonverbal behaviors. These interpretations of the client are in part formed by the counselor's awareness of psychosocial possibilities and through inferences from what the client says and does. Echoing Rogers' notion that symbolization of the phenomenal field is a working hypothesis, Weil describes the process this way:

Such observation should be treated as working hypotheses. Any particular reaction or communication may have alternate personal, social, or cultural explanations. As the session unfolds and as interventions based on the genetic counselor's inferences provide further information relevant to the assessment, the hypothesis can be further confirmed, revised, or rejected. Thus, there is an ongoing dynamic process of hypothesis generation, testing, and revision through which the genetic counselor obtains a better understanding of the counselee and refines her responses.[74]

The multiplicity of inputs that a counselor can use include: what, how, when, and why utterances are made and a host of nonverbal clues such as body posture and facial expression. The crucial challenge is tracking these inputs and at the same time providing appropriate outputs. Weil and Kessler assert that the ability to coordinate all of these activities comes with experience.

Whereas the teaching model categorizes psychosocial issues as an essential adjunct, the psychotherapeutic model places them at the center of its approach. This is especially true of emotional states. The Rogerian triad above paves the way to

---

[72] Jon Weil, *Psychosocial Genetic Counseling*, Oxford Monographs on Medical Genetics ; No. 41 (New York: Oxford University Press, 2000), 54. One significance of this quotation is that Weil motivates his approach to genetic counseling using an approach cited from an updated version of Roger's person-centered therapy.

[73] Ibid., 55.

[74] Ibid., 57.

develop and reduce emotional intensity as necessary.[75] If patients express strong emotion in the session, then it is critical to acknowledge these emotions in a sustained fashion. Weil gives the example of a patient who began a session in isolation and despair having called off his engagement thinking that he had Duchenne muscular dystrophy. The counselor explained to him that he had a much milder muscular disorder, and the patient grew angry. After an "empathic inquiry" into the sources of anger the patient expressed his despair and in follow-up sessions was able to accept his new diagnosis and see possibilities for renewing the relationship with his fiancé. By acknowledging the anger and asking questions about it, the counselor is recognizing its validity and providing opportunity for the counselee to talk about it. The crucial aspect of a counselor's recognition on Weil's view is that it seeks to understand the emotion from the counselee's perspective. This commitment requires sensitivity to the counselee's cultural background, which may encourage suppression of certain emotions.[76]

At other times the counselor must diffuse emotional intensity. In certain situations, the genetic counselor may discern that sustained attention on the emotional state of the counselee may overwhelm or produce anxiety in the counselee. In these circumstances, the counselor may in effect switch gears and discuss more cognitive content or utilize other coping mechanisms, e.g. humor, that are compatible with the defense mechanisms the counselee has already displayed. Weil acknowledges that this is a difficult task. A counselor must resist temptation to unconsciously adopt the counselee's defense mechanisms that seek to avoid emotions and at the same time use them circumspectly to diffuse emotions when necessary.

Promoting autonomy is at the center of the Rogerian model and is a key aspect in the psychotherapeutic approach to genetic counseling. Both Weil and Kessler assert that genetic counselors should aim at building a client's confidence in his or her ability to make good decisions. By explicitly supporting and praising a counselee's responses to present or past circumstances, the genetic counselor provides resources that can overcome feelings of being overwhelmed or inadequate. Weil states:

> Supportive statements of this sort are often unexpected which adds to their emotional impact. These comments help repair previous experiences of feeling judged, stigmatized, or treated with insufficient respect for autonomy. As with all reparative interventions, they may have a longer-term impact on self esteem and efficacy.[77]

This passage implies that counselees are often in need of such support and that encouraging words can actually be therapeutic in the sense of restoring the psyche to better functioning. Weil has confidence that the right communication techniques combined with the Rogerian triad of attitudes can successfully intervene in the psychological problems of the counselee.

---

[75] Ibid., 58–63.
[76] Ibid., 59.
[77] Ibid., 66.

A Therapeutic Vision of Communication   39

Much of what has been said thus far about the psychotherapeutic model depends on Weil's work and implicitly involves Kessler because he is cited throughout Weil's book. For the most part, Weil's telling of the story is endorsed in much of what Kessler has written. To rehearse Kessler's view on these same topics would be redundant. Instead, I turn to some of Kessler's early contributions that demonstrate how the psychotherapeutic model is informed by more global claims about communication.

Seymour Kessler has been articulating and updating the psychotherapeutic model of genetic counseling for over 30 years. In an early contribution he uses insights from theories of communication to inform the genetic counseling process. He offers several distinctions related to communication levels and context. Drawing primarily on theorists in psychotherapeutic process, he proposes that communication works at many levels and two are of particular significance: denotative and metacommunicative. The denotative points to the literal content of a communication that takes the form of a syntactically correct sentence. The meta-communicative picks out those aspects of communication below the level of consciousness that express needs, desires, and feelings usually in the form of nonverbal behaviors. One consequence of using this distinction is that incongruencies between what is said and the way it is said can be observed and responded to appropriately. The genetic counselor, who observes verbal compliance and nonverbal defiance, might circle back and address the issue in a way that allows the conflict to be resolved. Kessler shows how these levels of communication are affected by their contexts.

Kessler attends to the influence of context on communication by dividing it into three components: physical, social, and syntactic.[78] In terms of physical influences, a client comes to the counselor's "home ground" where the rooms, equipment, professionals and even smells have "stimuli value" for the client and implicitly encourage compliance.[79] Sensitivity to physical context clearly has implications for how a counselor might proceed to mitigate these factors in interactions. In terms of social features, Kessler wants us to see that a client's experience is influenced by whether the interaction is face-to-face, on the telephone, or between people of the same sex. Whether and to what degree a counselor is aware of this social dimension affects the education and counseling of a client. Syntactic context is the communicative circumstance in which a message occurs and the impact these circumstances have on meaning. Giving the example of a client repeating the question, "What are our chances of having a normal child?" Kessler points out that the question has a different meaning each time it is asked during the session. Recognizing syntactic context allows a genetic counselor to be sensitive to the movement of a conversation and the different roles an utterance can play at different times. When recognized, these con-

---

[78] Kessler's use of 'syntactic' rather than 'semantic' is questionable. Syntactic contexts usually refer to subsentential contexts and the rules that govern them whereas semantic contexts involve the circumstance in which a move is made in the language game.

[79] Seymour Kessler, *Genetic Counseling : Psychological Dimensions* (New York: Academic Press, 1979), 39.

textual factors can inform how a genetic counselor cares for the counselee throughout the conversation.

Kessler endorses the notion that "*all* messages contain requests" and that sympathy or caring is what is usually requested. Following Satir, he proposes that all messages seek validation on some level.[80] This commitment reinforces Peters' ascription that the therapeutic vision is a view of communication that extends beyond its disciplinary home toward a universal standing. If all acts of communication are requests for validation, then all communicators have the potential to validate or invalidate the other person. The psychotherapeutic model of genetic counseling seeks to validate the person through the aforementioned attitudes and techniques.

Evaluation

1. *Goals (a) to understand the other person (b) to bolster their inner sense of competence (c) to promote a greater sense of control over their lives (d) relieve psychological distress if possible (e) to support and possibly raise their self-esteem (f) to help them find solutions to specific problems* – The ambitious agenda of the psychotherapeutic model is both a strength and a weakness. In contrast to the teaching model, it has a more holistic view of caring for the counselee and offers many communicative techniques of which only a few have been mentioned. The main criticism offered here is: If the teaching model seeks to do to little for the patient, then the psychotherapeutic model aspires to do too much. Intended for situations where the counselor-counselee's relationship has time to develop over many sessions, the Rogerian approach is not appropriate for the institutional circumstances that define genetic counseling. Genetic counseling relationships are usually developed over one or two sessions that lasts anywhere from 30 min to 2 h. Kessler has acknowledged that the Rogerian approach must be modified in significant ways to fit the circumstances of genetic counseling.[81] Kessler's solution has been to emphasize the need for counseling skills that are customized for short-term interactions: "This might require the acquisition and development of rapid means of assessing others and understanding their needs, the skills seasoned professionals tend to develop in any case."[82] In light of this revision, attaining the goals above becomes more dependent on effective techniques that allow quick evaluations than on the relational benefits conferred by the Rogerian triad of attitudes. Rogers himself would question whether this technique-driven approach can obtain the substantive goals above.

---

[80] Ibid., 40.

[81] S. Kessler, "Psychological Aspects of Genetic Counseling. Xi. Nondirectiveness Revisited," *Am J Med Genet* 72, no. 2 (1997): 166. His stance has evolved over the years and can be interpreted as slowly revising the Rogerian approach.

[82] Kessler, "Psychological Aspects of Genetic Counseling. Ix. Teaching and Counseling," 293.

The question becomes whether the psychotherapeutic model in these institutional circumstances can hold up under the weight of its goals and tasks given its acknowledgment of the complexity of human psychology. Rapid evaluations of complex phenomena is a difficult trick to pull off especially in the psychosocial realm. As Kessler asserted at the beginning, it takes an exceptional individual to combine the teaching and counseling models. For example, the first goal of understanding the client is questionable in terms of not only the temporal constraints but also epistemic ones. A comprehensive one-way understanding of another person – not to mention mutual understanding – is difficult to achieve even with ample time and access. Awareness of this difficulty is seen in the repeated disclaimers of those who endorse this goal.[83] Advocates of empathic inquiry admit how difficult it is to bracket all evaluative background commitments and to bridge actual differences in perspectives particularly when cultural backgrounds are significantly far apart. The constraints of genetic counseling thwart an already difficult epistemic aim. If all the other goals flow through this first one, as Kessler claims, then they are all put into question by these obstacles.

2. *Based on perception that clients come for counseling for complex reasons* – On the one hand, this commitment is a responsible stance given that patients (1) some times do not know that they have come for genetic counseling (2) have different perspectives that generate different needs (3) sort through genetic information over time as their needs change within and between sessions. On the other hand, this thesis fails to articulate the assumptions about patient motivations that are necessary to justify saying anything at all. In the next chapter, I will propose that assumptions have to be made about patient motivations to justify the expectation that genetic counseling might help the patient.

3. *The model has complex assumptions about human behavior and psychology which are brought to bear in counseling* – The expectation that genetic counselors should have a complex grasp of human behavior and psychology raises several questions. First, how much knowledge of psychology is sufficient to perform genetic counseling? Second, since there are competing accounts of human behavior and psychology, which one should genetic counselors endorse? The third and possibly most important question is: How should a complex theory be used in a session?

The Rogerian vision of psychotherapy attempts to bracket substantive psychological assumptions to avoid imitating the diagnose-and-treat model of traditional medicine. This method is committed to the therapeutic value of communication accompanied by the appropriate attitudes. The difficulty in undertaking this kind of communication is that empathic identification requires a bracketing of psychological assumptions to understand the person from her perspective, a perspective that presumably lacks a sophisticated psychological theory. The genetic counselor taking such a stance must make inferences from what he hears and sees. From what set

---

[83] Carl Rogers, Jon Weil, and Seymour Kessler all attend to the limitations of aiming towards empathic identifcation.

of commitments are these inferences made? Propositions from a psychological theory would be likely candidates as premises to understand and respond to observations and yet the use of such commitments can conflict with the original empathic stance.

An example helps illustrate this point. A counselor observes that a 25-year-old client whose child has been diagnosed with Down syndrome is feeling guilty about giving birth to a child with cognitive limitations. What should the genetic counselor infer from this observation? A fully empathic stance would try to understand the patient's experiences and beliefs that produce the guilt and grasp the consequences of this guilt for the patient. A responsibility of achieving this level of understanding within a Rogerian model is to mirror the patient's perspective back to her without evaluation. The goal of this process is that the patient will recognize incongruencies between the mirrored perspective and her true self. The true self is the final arbiter.

Leaving aside questions about what 'true self' means, this empathic stance is highly improbable given that the psychotherapeutic model is generally skeptical that a patient's guilt is warranted in circumstances similar to the one described above. A more likely response to a patient's guilt in the psychotherapeutic model is to infer that this guilt is unjustified and should be alleviated.[84] The genetic counselor states authoritatively, as a representative of the scientific community, that the chromosomal abnormality is a random occurrence and not the fault of any individual agent. Has any harm been done? The statement might make the client feel less guilty, but it could make her question whether God is in control. This doubt could undermine her hope that the Lord will provide the strength to deal with this new circumstance. This outcome is not inevitable. The counselor could help the counselee reconfigure an understanding of God's sovereignty that allows for randomness and hopefulness. This result does not avoid the collision between an empathic identification and a substantive stance on guilt and causality. In this case, a psychological assumption about guilt and a biomedical conclusion about randomness influence the revision of a patient's religious beliefs. This case does not challenge the claim that genetic counselors should have complex views about patient behavior and psychology. It only highlights the complexities of adjudicating psychological assumptions about the efficacy of empathy with other substantive commitments about appropriate psychological states.

4. *Counseling task complex: (a) requires assessment of client's strengths and limitations, needs, values and decision trends (b) requires a range of counseling skills to achieve goals and (c) requires individualized counseling style to fit client's needs and agendas; flexibility (d) requires counselor to attend to and take care of his own inner life* – If the goals of the psychotherapeutic model underestimate discursive challenges, then the tasks developed to meet those goals are vulnerable to the same charge. The tasks in (a) require the genetic counselor to make a number of complex attributions in a very short period of time. Should a

---

[84] Weil, 20–21. Weil characterizes some patients and their guilt as substituting personal responsibility for the existential void of randomness in order to avoid the reality of contingency.

genetic counselor be expected to reach such conclusions – even provisional ones – about a counselee?[85] The tasks in (b) should be endorsed because most items on the list should also be on a list for effective communication. Informed observation, developing and diffusing emotional states, reframing situations to name a few are skills that can help a client grasp the meaning of the genetic information. Employing these skills effectively does not require a complex set of psychological assumptions; nor must they serve the psychological goals. I will demonstrate how they can function within an explicitly normative framework in Chap. 3. The value of (c) is that it guides counselors away from treating all counselees the same, e.g. as idealized rational persons. The (c) tasks can be interpreted in a strong sense where the genetic counselor has a comprehensive grasp of the client's perspective and can confidently counsel him or her with this knowledge. A weaker stance, the one I think more defensible, is that the HCP must be responsive to what the individual says and does in the session as they coordinate the meaning of the information together. The important difference between the strong and weak stance is that the clinician acknowledges the provisional status of his or her understanding of the client. The final task is a worthwhile undertaking that is reminiscent of living Socrates' examined life. The importance of this task for this project is that it requires attention to the robust perspective that a genetic counselor brings into the session. The wording "inner life" utilizes the spiritualist tradition's picture of inner/ outer and attenuates the scope of commitments that actually need attention. The genetic counselor should not only be aware of his or her emotional habits, personal beliefs and history of actions but also the web of institutional norms that are formed by the historical tradition of biomedical practice. Whether these are part of her "inner life" does not matter as much as the awareness that these institutional and other role-based norms are operational when he or she inhabits the role of genetic counselor.[86]

5. *Education is used as a means to achieve above goals.* All of the goals of the psychotherapeutic model are psychological according to Kessler.[87] Education should facilitate understanding between persons, promote inner competence, provide control, relieve distress, raise self-esteem and enhance problem solving. The details of the relation between education and these goals is not fully worked out by Kessler. Two critical question need to be answered. How does education actually serve the above goals? Should education be understood as a means to these ends?

Thinking about Debbie's case will help us answer the first question. When the genetic counselor explains to Debbie her risk status, the above goals should be served in some way. If Debbie grasps the information, then she could experience the

---

[85] What I hope to show in the next chapter is that these kinds of assessments can be facilitated by the HCP without taking on the responsibility of actually making judgments about clients.

[86] E. W. Clayton, "The Web of Relations: Thinking About Physicians and Patients," *Yale J Health Policy Law Ethics* 6, no. 2 (2006): 472–75. Clayton provides a helpful synopsis of the competing interests and professional shortcomings that affect even the most thoughtful of physicians.

[87] Kessler, "Psychological Aspects of Genetic Counseling. Ix. Teaching and Counseling," 290.

empowerment of understanding her risks. This empowerment comes from knowing her pregnancy status and having choices to do something about her risk. Instead of going through her pregnancy anxious and ignorant of important facts, she now knows about her option to find out for sure whether her baby has certain abnormalities. With this information she can plan for the future and feel as though she has taken full responsibility for her pregnancy. Education can have this effect on Debbie's psychological state and the genetic counselor can contribute to this effect. The difficulty with this picture of educational consequences is that a very different set of consequence are also likely to occur. Debbie could also have a very different experience of learning about her risk status. The information could shake her faith and produce an intense anxiety about the pregnancy that she would have never experienced until learning about the information. The experience of a dilemma, which is described in the case, is in part caused by the educational process. The emotion and ambivalence that Debbie experiences can last a long time after the genetic counseling session is over. Kessler and Weil would acknowledge these possibilities and rejoin that with the appropriate counseling skills the genetic counselor could mitigate the negative effects on Debbie's psychological state. This response is plausible but remains distant from the characterization that education is a means to achieving ambitious psychological goals.

How education affects the patient depends on the circumstances of the patient and the skills of the counselor. Because of these contingencies, education should not be seen as a means to the above psychological goals. Instead, it should be characterized as having significant influence on the feasibility of the goals above. A crucial skill that a genetic counselor must have is recognizing what impact education does have on a patient. This primarily comes from listening to what the patient says and observing nonverbal communication. Education may actually thwart the above goals in an enduring way and it is vital that the genetic counselor recognize this change in the patient's status.

6. *Relationship aims for mutuality* – Kessler's model instructs a genetic counselor to de-emphasize the authoritative structure of the relationship without relinquishing it.[88] Efforts to downplay the authoritative structure entail letting the counselee steer the conversation when appropriate, championing the autonomy of the patient and demonstrating a level of empathy that expresses, "I'll be there with you in case you stumble."[89] These measures all aim towards mutuality. Or do they? Kessler never defines what he means by mutuality, and Weil does not use the term to define the contours of his project. For this thesis to become plausible, some definition of mutuality needs to be given. If it refers to the aim of mutual *understanding*, then this thesis becomes tenuous because of the difficulty of achieving anything close to a reciprocated grasp of the situation. Kessler's description of genetic counseling cited in the introduction implies that this might be a worthy goal: "Their assumptions about things seem vastly different and

---

[88] Ibid., 291.
[89] Ibid.

there are other impediments to communication and mutual understanding."[90] Such an aim fails to take the differences in perspectives seriously.

If mutuality refers to the mutual *recognition* of persons as distinct sites of authority, then the thesis becomes more plausible. Kessler recognizes the imbalance of authority in the professional-client relationship and the failure of clients to recognize their own authority in the situation. If the psychotherapeutic model adopted the second definition of mutuality, then it could become the unifying concept of the model whose central thrust is to promote client autonomy. This unification puts into question the need for an ambitious set of psychological goals and assumptions that currently define the psychotherapeutic approach. For example, the first goal of the model is to understand the other person through empathic identification. The need for rigorous bracketing of the professional perspective is not necessary to encourage the client to acknowledge his or her own authority in the interaction. The call for emptying one's perspective to absorb another's is a clear remnant of the spiritualist desire to share interiors and an application of Roger's therapeutic vision of communication. What is needed is a competency in eliciting client participation in a dialogical process with the HCP. This alternative will be developed in Chap. 3.

Summary

Peters tells an expansive story about communication that has as its central conflict the frustrated ideal of sharing human interiors. We should be unified but we are divided. He identifies the technical and therapeutic visions as initiatives to make the dream of shared interiors a reality. Through precise, discursive transmissions or genuine, empathic identifications, the reality of possessing the identical information or the possibility of mutual understanding seems to appear on the horizon. The central claim of this chapter is that the teaching and psychotherapeutic models of genetic counseling are inheritors of the spiritualist tradition and respectively employ the theses of technical and therapeutic visions of communication. Peters is right to reject the spiritualist tradition's interpretation of our discursive condition and to criticize the communication visions that it has generated. An alternative tradition is offered in the next chapter as well as a theoretical foil to the two visions rehearsed above. These expressive resources will locate and underwrite the responsibility model of genetic counseling.

---

[90] Kessler, "Psychological Aspects of Genetic Counseling: Xii. More on Counseling Skills," 263.

Chapter 3
A Responsibility Model of Genetic Counseling

> If I believe that Zoroaster is the sun and that its shining is his beatitude, then an utterance of 'The sun is shining' means something different in my mouth than it does in your ears."
> (Robert Brandom, *Making It Explicit*)

In the previous chapter, I introduced and elaborated the teaching and psychotherapeutic models of genetic counseling. The models were located within larger narratives from which they inherited distinct views of communication. I claim that the teaching model of genetic counseling is underwritten by the technical vision of communication; the psychotherapeutic model by the therapeutic vision of communication. Both are broadly situated in what John Durham Peters' calls the spiritualist tradition. In this chapter I develop an alternative view that has been offered in the genetic counseling literature that places responsibility at the center of its approach. After introducing the responsibility model, I elaborate it in reference to a different philosophical story about communication whose details are worked out in Robert Brandom's pragmatic model of conversational scorekeeping. These additional expressive resources are then used to flesh out the theses of the responsibility model.

Responsibility Model

If the primary goal of the teaching model is help the counselee understand and the central goal of psychotherapeutic model is to help the counselee adapt, then the goal of the responsibility model is to help the counselee take responsibility for receiving genetic information. The genetic counselor as teacher aims to impartially transfer genetic information to the client who will then possess balanced information to make a "rational" decision. The genetic counselor as psychotherapist intends to empathically understand the counselee and intervene as appropriate to promote the autonomy of the individual. As a responsible communicator, the genetic counselor attempts to coordinate meaning across different perspectives with the goal of helping the counselee grasp and make decisions with the genetic information. This alternate frame for understanding genetic counseling is by no means unprecedented.

Mary White criticizes nondirective models of genetic counseling[1] and proposes an alternative model called "dialogical counseling" that seeks to promote responsible decision making.[2] Nondirective counseling, in White's terms, is defined by a negative right that entitles counselees' decision-making processes to be protected from interference or coercion. Upholding this right restricts what interventions the counselor can undertake. Once the genetic information has been carefully explained and all pertinent questions answered, the genetic counselor must remove possible constraints on the autonomous decision making process of clients. White challenges the understanding of autonomy that is presupposed in nondirective counseling. If autonomous decision making means deliberation that is free from constraints, then nondirective counseling only takes into account external constraints and fails to acknowledge internal constraints to autonomy such as misconceptions or stress. With these critical insights in mind, White suggests an approach to autonomy and counseling that is informed by notions of sociality and responsibility.

In constructing a counseling model around social responsibility, White utilizes the work of protestant theologian, H. Richard Niebuhr. Drawing primarily from Niebuhr's *The Responsible Self*, White promotes an understanding of autonomy that takes into account the sociality and interdependence of human beings. Niebuhr claims that selves are social, practical phenomena who come into existence through recognitive relations with other selves. If this account of humanity is true, then autonomy cannot be an atomistic capacity of self-determination. Autonomy and more specifically deliberation and decision making must be seen as social phenomena that involve interactions such as dialogue. White points out that dialogue with people we trust is a common form of deliberation; and if decisions are made individually, then they often involve an internal dialogue. In the interest of noninterference, the nondirective counseling model limits the counselor's role in deliberation leaving the counselee to her own devices to make a decision in unfamiliar discursive territory. White claims that this is a breach of social responsibility, a failure to recognize the counselor's responsibilities in an important dialogue.

Niebuhr's notion of social responsibility is specified in reference to the goals and strategies of genetic counseling. In the dialogical model, responsible decision making is the goal of genetic counseling and dialogue is the strategy that genetic counselors undertake to achieve it. Following Niebuhr, responsible decisions bear the marks of a dialogical process where: (1) every action is a response to a prior action (2) all actions – in contrast to behaviors – involve interpretation of what is happening (3) an agent anticipates consequences of possible actions (4) fitting actions are

---

[1] M. T. White, ""Respect for Autonomy" In Genetic Counseling: An Analysis and a Proposal," *J Genet Couns* 6, no. 3 (1997): 298–99. White acknowledges that nondirectiveness has many meanings. Her criticisms are most consistently applied to a version of nondirective counseling that resembles the teaching model presented in Chap. 2.

[2] Ibid., 304.

acknowledgments of ongoing individual and collective narratives. In short every responsible decision asks: "To whom or what am I responsible and in what community of interaction am I myself?"[3] The offering of genetic information is an action undertaken by the genetic counselor and becomes the prior action to which a counselee must respond. In terms of normative trajectories, they share responsibility for the genetic information like travelers on different paths share a crossroad.

Genetic counselors have responsibilities to their institutional and larger social contexts as well as to the context of the client whose background commitments represent a more or less different repertoire of responsibilities. A genetic counselor who enacts social responsibility must try to acknowledge all of these contexts in the offering of the genetic information. White illustrates these responsibilities in the area of prenatal diagnosis. She notes that the practice of medicine has values, albeit contested ones, that define the circumstances under which it is appropriate to use prenatal diagnosis.[4] These professional standards serve both as constraints in any genetic counseling session and as the perimeter for permissible speech. Harmful effects of Down syndrome on fetuses can be discussed whereas the harms of being male or female cannot be. Responsibilities to social solidarity, White admits, are much more difficult to discern:

> Thus in contemporary secular society social solidarity may best be conceived as embracing a plurality of evolving social and cultural values in which decisions may be considered ethically responsible each corresponding to different moral priorities.[5]

Genetic counseling in prenatal circumstances has prompted diverse responses to the status of terminating a pregnancy after a diagnosis of Down syndrome. By defining social solidarity in reference to pluralism, the genetic counselor is in a better position to help individuals take responsibility for the information in reference to his or her perspective and cultural values that shape it. In Debbie's case, the genetic counselor shares responsibility for helping her respond to emotional, moral and religious issues. At the same time, the contract for sharing such responsibilities has to be negotiated with the client. In other words, an HCP cannot assume that patients want the same level or kinds of help.

Many of the contents of White's proposal are either taken over or modified in the responsibility model presented here. In the interest of comparing like models, I use Kessler's schematic form to introduce the responsibility model:

1. Goal: (a) To help coordinate the meanings of genetic information across diverse perspectives (b) To facilitate responsible decision making
2. Based on assumption that clients come to share responsibility for understanding the genetic information and for decision making

---

[3] H. Richard Niebuhr, *The Responsible Self; an Essay in Christian Moral Philosophy*, [1st] ed. (New York,: Harper & Row, 1963), 68.

[4] As White acknowledges, this is an area of heated debate. For an extended discussion of these issues see Parens and Asch.

[5] M. T. White, "Making Responsible Decisions. An Interpretive Ethic for Genetic Decisionmaking," *Hastings Cent Rep* 29, no. 1 (1999): 18.

3. The model assumes that the patient can participate in a dialogical process of grasping the genetic information and making responsible decisions.
4. Counseling task is to facilitate (1) navigation and negotiation of the appropriate perspectives for understanding the genetic information and (2) practical reasoning about what action to take in reference to the relevant sources of responsibility
5. Relationship aims towards the mutual recognition of shared responsibilities

By designating responsibility as the conceptual hub, the responsibility model transposes most aspects of genetic counseling into an explicitly normative key. For example, taking a pedigree involves giving a person a genetic identity, an identity that has to be negotiated with other ways of sorting the self such as a child of God, successful professional, or dutiful citizen.[6] Uttering genetic information to a patient has a two-fold normative significance: (1) The patient receives a true claim about her genetic situation that she *should* add to her repertoire of commitments. (2) The patient *should* use this knowledge in a way that will determine a course of action. In the responsibility model, the perspectives of both counselor and patient come into focus as sites of normativity that must interface in the process of interpretation and decision making.

As it is used here, the concept of normativity extends beyond norms that guide practical intentions to include interpretive norms that inform the ongoing formation of beliefs. The sphere of the normative refers to a social practical space where moves, both linguistic and nonlinguistic can come under appraisal. In other words, normative moves are ones that can be attributed as wrong, inappropriate, or unfitting. One lesson Kant taught is that a judgment, e.g., "That banana is yellow" or "Your risk is such and such," is the most basic unit of responsibility.[7] According to Hegel the resources for appraising a judgment are fundamentally social and historical. Being responsible to shared rules of inference is one way to think about this kind of responsibility.

If discursive resources vary within a pluralistic society, then the meanings within the clinic will be coordinated from several standpoints. For example, risk assessments of pregnancies can be appraised in relation to several different contexts. In the context of the rules for probable reasoning, which correlate blood levels with probabilities of Down syndrome, a risk assessment can be judged incorrect. Without specialized training, patients are not in position to appraise the correctness of the probability but they are competent to make other kinds of appraisals. If they are associated with a religious community that prohibits abortion under any

---

[6] D. Armstrong, S. Michie, and T. Marteau, "Revealed Identity: A Study of the Process of Genetic Counselling," *Soc Sci Med* 47, no. 11 (1998): 1653–8. Arrmstrong and others report how genetic counseling constructs a genetic identity.

[7] Immanuel Kant, "An Answer to the Question: What Is Enlightenment?," in *Kant: Political Writings*, ed. Hans Siegbert Reiss (Cambridge ; New York: Cambridge University Press, 1991). Robert Brandom asserts that one of Kant's most important insights is the relationship between judgment and normativity. Brandom draws a sharp contrast between Kant's and Descartes' accounts of judgment.

circumstance, then the patient might conclude that the risk assessment and amniocentesis are inappropriate because they endanger the fetus for no justifiable benefit.[8] These judgments are normative because they are responsible to sets of shared commitments from which correct and incorrect inferences can be made. It is within this understanding of the normative realm that White's commitment to social responsibility should be understood.

If genetic counseling's goals, assumptions, tasks and overarching aims are to be understood in explicitly normative terms, then a development of normativity and responsibility is required and will be pursued in the next two sections of this chapter.[9] The normative elements as well as the other components of the model are related to an alternative tradition of spirit and a pragmatic theory of communication. The theses of the responsibility model of genetic counseling are understood in reference to a "more robust tradition of spirit"[10] and underwritten with an account of communication set out by Robert Brandom. The first move situates the responsibility model within a largely Hegelian picture that develops the problems of spirit in a way that motivates the second move. Brandom's project, which explicitly acknowledges its Hegelian roots, works at a finer level of detail to see how normativity, authority and responsibility are at work in the interpretation and understanding of utterances. Through these efforts, I aim to locate and underwrite the responsibility model of genetic counseling within a larger philosophical picture.

Embodiment Tradition of Communication

Peters claims that the spiritualist tradition and its concomitant visions of communication fail to develop the "pragmatic middle ground of making due."[11] The terrain that needs cultivating is a complex attitude toward communication that resists aspirations for complete identity with each other and at the same time avoids resignation to the fate of discursive distance. Undertaking the former potentially overrides differences and results in the replication of selves or groups; endorsing the latter leads to rash claims of incommensurability and isolation. Two conclusions from the last chapter are significant here. The psychotherapeutic model of genetic counseling aspires for too much identification in its goal of empathically understanding the client. The teaching model, to overcome or avoid imperfect media of subjectivity,

---

[8] Whether knowing a fetus has a condition confers benefit to a pregnant woman that is not considering termination is a real question. I observed two OB/GYNS who had different stances on the issue. One stated that if a woman would not consider termination then, she should not undergo amniocentesis. The other said to patients that knowing whether a baby has Down syndrome can help parents and health care provider prepare for the delivery.

[9] The debate about the status of norms as regularities or proprieties cannot be sorted through in the confines of the project. The position that norms are proprieties is the one undertaken here and depends largely on Robert Brandom's work.

[10] Peters, *Speaking into the Air: A History of the Idea of Communication*, 109.

[11] Ibid., 65.

places its confidence in too narrow a set of transmissible messages that involve primarily objective information. The pragmatic middle ground developed in this chapter articulates communication, more specifically linguistic practice, as a normative and social achievement that mitigates differences without aspiring to erase them and that allows individuals to coordinate a world through an expanding range of meanings.

The spiritualist tradition, introduced in Chap. 2, claims that spirit – the mental or the normative – resides in our individual interiors and that exteriors, i.e. speech – producing bodies, obstruct the sharing of individual spirits. Its assertion of an autonomous mental sphere gives ideas and intention primary roles in a metaphysics of intentionality and has supported a variety of accounts of communication from Plato's recollection, to Augustine and Aquinas' angelology and to Locke's notion that ideas excite identical mental states. In this dualistic picture, exteriors, i.e. language and bodies, are different *kinds* of things than interiors, i.e. ideas. Exteriors produce sounds and scripts that are imperfect mediums seeking to bridge separated and isolated interiors. Two key problems get articulated by this picture: (1) Individual interiors once united are now separated. (2) Available mediums that link interiors fall short of the task of uniting them. The problems that define this tradition, as Peters tells the story, are revised or rejected by another tradition of thought whose origin is marked in the writings and legacy of Hegel.

Peters' narrative highlights the development of a social and practical account of the relation between spirit and communication. If the spiritualist tradition asks how interiors can be reunited, then this alternative tradition asks how human bodies interact to give rise to concepts, concept mongers and coordinated worlds. Hegel's enduring answer to this question is that this is a social-historical process, which he calls Spirit or *Geist*, that begins with reciprocal recognition:

> For Hegel communication is not a psychological task of putting two minds *en rapport* but a political and historical problem of establishing conditions under which the mutual recognition of self-conscious individuals is possible. The issue is to reconcile subjects with their embodied relation to the world, with themselves and with each other.[12]

The problems of communication on this view can be understood in reference to several factors. One of the key insights of Hegel is that we are not transparent to ourselves. Selves do not begin as private spirits or interiors encased within a frustrating material home; instead selves are wrought through social practices, e.g. linguistic practice, where public meanings and other skills are generated by specific kinds of interactions between individual bodies. Hegel uses the example of a master and slave to illustrate this point.[13] The uptake is that we need others to be self-conscious and this dependency creates the possibility of constructing others in our own image or being constructed in the image others.

---

[12] Ibid., 112.
[13] Georg Wilhelm Friedrich Hegel, Arnold Vincent Miller, and J. N. Findlay, *Phenomenology of Spirit* (Oxford: Clarendon Press, 1977), 111–19.

If the account of selves as socially constituted is endorsed, then it has the practical implication of making dialogue a condition for the existence of self-conscious selves. Dialogue is the reciprocal practice of coordinating meanings across perspectives that allows us to take explicit responsibility for and be responsible to our selves, each other and our world. White's notion of dialogical counseling draws its theoretical momentum not only from Niebuhr's conception of responsibility but also implicitly from Hegel's notion of reciprocal recognition. But all communication is not dialogical as Peters is quick to point.

Dialogue must be tempered by placing some value on dissemination. On Peters' view, dissemination is an act of communication that cannot or does not concern itself with reciprocation. Mass communication is one example and Jesus' parables are another.[14] In the former, dialogue is logistically improbable; in the latter, it is eschatalogically unworkable. Peters' attention to this dimension of communication is important for this project for several reasons. On the one hand, it demonstrates the kind of communication genetic counseling should *not* be under most circumstances. A genetic counselor who casts seeds of genetic information to patients with the attitude of Jesus' sower should be criticized for a lack of care. On the other hand, dissemination can be an appropriate stance in an important set of circumstances that occur in genetic counseling: the moments when patients do not want to talk. HCPs can prompt but cannot ultimately make the patient participate in a dialogue. It does not take much imagination to see the harms of trying to force reciprocal communication in these situations. Peters calls advocates of dialogue who do not appropriately recognize these limitations, "dialogians."[15]

This alternative understanding of communication suggests a different story about what is happening in the 'Patient Education' room as Debbie and the genetic counselor try to understand one another. The picture of two interiorities separated by imperfect media is replaced by two embodied and distinct perspectives trying to navigate a shared world with available communicative resources. The possibility of communication requires that Debbie and the genetic counselor share many common beliefs and concepts. Both assume, for example, that the other has some very basic knowledge about what 'pregnancy,' 'risk,' 'needle', and 'termination' mean. Such concepts are common references in their shared world. At the same time the need to communicate arises from what they do not share. It is assumed by both of them that the genetic counselor possesses medical information about Debbie that she does not yet have. This mutual assumption implicitly reflects a recognition of differences in their individual perspectives. It also suggests a tacit awareness of the cultural phenomena of specialization that systematically develops different perspectives in the training of individuals to undertake specific roles such as that of a genetic counselor. Communication can be understood as the capacity to navigate meanings between perspectives against this backdrop of assumed commonalities and differences. Will this information be helpful to Debbie? Does Debbie understand probabilities or Down syndrome? Is pregnancy termination permissible to her? Why is the genetic

---

[14] Peters, *Speaking into the Air: A History of the Idea of Communication*, 51, 206.
[15] Ibid., 34.d

counselor giving Debbie this information? Does the genetic counselor think that fetuses with Down syndrome should be terminated? As these two perspectives enter the conversation, they must try to coordinate their distinct standpoints within a shared but specialized world that can produce specific facts about the embodiment of Debbie and her fetus.

Peters' Hegelian story, which he develops further with Kierkegaard and Marx, offers a trajectory of thought about communication that poses a different set of problems than the spiritualist tradition. This alternative view asks: *How do we as bodily creatures coordinate ourselves in a way that allows us to build worlds together from individuated standpoints?* Language is clearly one of the most important resources in this development and Hegel's robust vision of spirit continues to press twenty-first century thinkers for a more detailed understanding of language and communication. In the next section, a pragmatic theory of communication is offered that reflects the insights of this tradition as it has taken shape in the work of Robert Brandom. My aim in introducing his work is to provide expressive resources for underwriting the responsibility model of genetic counseling.

A Pragmatic Theory of Communication

One of central questions of this project is: What understandings of communication are and should be operational in genetic counseling models? In Chap. 2, I tried to show the visions of communication that are implicit and at times explicit in two dominant models in genetic counseling. Brief critiques of these models and their respective visions have been offered. Having introduced the responsibility model and a framework of core concepts that fund it, a more detailed theory of communication is needed to show how normativity, authority and responsibility are operational in a fundamental feature of genetic counseling, the exchange of utterances.

The methodological strategy underway assumes that a grasp of genetic counseling should begin with a general account of the possibilities and constraints that constitute the practice of communication. The overarching features of communication must be understood before giving an account of how competencies specific to health care professionals and genetic concepts further constrain or specify genetic counseling. One could characterize this methodological sequence as Aristotelian in its movement from a less qualified description of an activity or object, e.g. Aristotle's treatment of friendship, to more qualified accounts, e.g. Aristotle's account of pleasure-based, utility-based, and virtue-based friendships. In providing a theoretical account of the basic features of communication such as attributing, acknowledging, inferring and understanding, a new set of expressive resource is offered for reflection on the challenges in genetic counseling. A brief hypothetical example of a theoretical benefit might prove helpful.

If a genetic counselor implicitly holds, as the therapeutic vision proposes, that empathic understanding is the ideal of all communication, then she would experience consistent frustration in the outcomes of genetic counseling. This frustration might compel her to ask more and more psychosocial questions and to provide more elaborate psychosocial interpretations of the situation. Pressing towards empathic understanding, the genetic counselor ignores signs that the patient is uncomfortable with talking about these personal matters. As a result the genetic counselor institutes what Peters' has called the "tyranny of dialogue"[16] an overriding concern for comprehensive and equal exchange that ignores the needs of, in this case, the patient. If the same counselor took communication to be a coordination of meanings across different perspectives, then her responsiveness to the respective backgrounds should be keener and her expectations for outcomes should be recalibrated in reference to the intractability of some differences between perspectives. Although this hypothetical is extreme, it shows that attitudes, implicit or explicit, towards communication can affect how genetic counseling is undertaken.

Robert Brandom's[17] model of deontic scorekeeping articulates the structures of rudimentary conversational exchange and offers this project a vocabulary and system to think about the responsibility model. The next two chapters elaborate the consequences of the responsibility model and this pragmatic theory of communication in relation to concerns about nondirectiveness in genetic counseling and the complex issues surrounding religion. In an effort to stay close to the practical question, a likely utterance from Debbie's case is used to help explicate the model. Thus, an ancillary goal will be to understand what it might mean for the following sentence to travel from the genetic counselor's mouth to Debbie's ear: *You have a 1/106 risk of giving birth to a child with Down syndrome.*

What Is Communication?

Brandom's *Making It Explicit* develops and works out a theory of language use that he terms deontic scorekeeping.[18] Communication is the joint practice of deontic scorekeeping between two or more scorekeepers. The core of this practice is giving

---

[16] Ibid., 159.

[17] In philosophical circles, the choice of Brandom over Jurgen Habermas might be scrutinized. Two reasons justify this decision. First, Habermas's work focuses more on communicative norms within procedural contexts of 'ideal speech', whereas Brandom's applies to a broader range of communicative contexts. Second and more important, Brandom's account supplies the details to Habermas's pragmatic stance. Habermas recounts a letter he received from Richard Rorty that recommended Brandom's work as working out the pragmatics of communication that Habermas intends. For reference to letter, see Jürgen Habermas, Ciaran Cronin, and Max Pensky, *Time of Transitions* (Cambridge, UK ; Malden, MA: Polity, 2006). and for a helpful comparison of Brandom and Habermas, see Kevin Scharp, "Communication and Content: Circumstances and Consequences of the Habermas-Brandom Debate," *International Journal of Philosophical Studies* 11, no. 1 (2003): 43–61.

[18] Every theory has limits and this one is no exception. As Brandom indicates, his theory is an "artificial idealization" that oversimplifies and schematizes what we do but at the same time we should be able to recognize our own linguistic practices in this account. Brandom correlates his term 'commitment' with 'obligation' and 'entitlement' with 'permission' and explains his resis-

and asking for reasons.[19] This section does not entail an exhaustive elaboration of Brandom's model, which he works out in a 740-page book, but presents his theory in enough detail to underwrite the responsibility model. Deontic scorekeeping has two wheels on which it rolls: normative pragmatics and inferential semantics. Brandom encapsulates the basic structure of the model in this passage:

> Competent linguistic practitioners keep track of their own and each other's commitments and entitlements. They are (we are) deontic scorekeepers. Speech acts, paradigmatically assertions, alter the deontic score; they change what commitments and entitlements it is appropriate to attribute not only to the one producing the speech act but also to those to who it is addressed. The job of pragmatic theory is to explain the significance of various sorts of speech acts in terms of practical proprieties governing the keeping of deontic score –what moves are appropriate given a certain score and what difference those moves make to that score. The job of semantic theory is to develop a notion of the contents of discursive commitments (and the performances that express them) that combines with the account of the significance of different kinds of speech act to determine a scorekeeping kinematics.[20]

To develop a pragmatics without yet dipping into semantics, Brandom asks us to imagine observing social practices where performers make nonlinguistic moves that have the significance of changing their normative (and social) status. The two basic statuses in play are commitments and entitlements.

Brandom makes a fundamental distinction between entitlements and commitments.[21] *Entitlements* involve attributions of authority to do something that cannot be done without such authorization. Brandom gives the example of the ticket taker at a movie theatre who authorizes those with tickets to enter. This practice does not commit the ticket holder to enter but does allow non-ticket holders to be ejected, the removal being understood as a type of sanction. Brandom construes this example of ticket taking as attributing authority without yet attributing responsibility. A person is permitted without yet being committed. Medical referrals can function as tickets or entitlements to other kinds of medical services and in the same way only authorize but do not commit a patient to those services.

The structure of *commitment* involves responsibility and authority. Brandom uses the example of "taking the queen's shilling," an eighteenth century British practice where taking a shilling from an official recruiting officer meant that the

---

tance to these correlates as disrupting the picture that authority necessarily depends on hierarchy. Commitments and entitlements are normative statuses for Brandom and one of his central contributions is the development of these normative statuses in reference to assertional responsibility. One possible consequence of his development is that ethical theories must begin at the level of the norms of discursive practice.

[19] Brandom acknowledges that placing the practice of giving and asking for reasons at the core of linguistic practice is an intellectualist move away from Wittgenstein's notion that language has no downtown. Wittgenstein insight is that we use language for all kinds of purposes and thus there is no core or 'downtown' use for language. Brandom's insight is that linguistic practice cannot be *understood* at all without the practice of giving and asking for reasons.

[20] Robert Brandom, *Making It Explicit: Reasoning, Representing, and Discursive Commitment* (Cambridge, Mass.: Harvard University Press, 1994), 142.

[21] Ibid., 159–62.

A Pragmatic Theory of Communication   57

taker was expected to serve in the military.[22] The actual taking of the coin changes the taker's score, his normative status, by making it appropriate for others to attribute to him a commitment to serve the military. The consequence of being attributed this commitment consisted of undergoing sanctions levied in circumstances where there was failure to serve. Two features of commitment can be understood from this practice. First, to undertake a commitment involves a performance that signifies to others that they can attribute a commitment or responsibility to the performer. Second, attributing commitments consists in whatever taking someone as committed entails either defined externally by an observer or internally by the practice. In the case of the queen's shilling, failure to fulfill the commitment involves the attributors being disposed to sanction, i.e. a beating or court martial. Beating is normatively less complex than the processes of court martialing.[23]

Brandom wants us to see the possibility of more sophisticated rendering of this practice by extending its specification in terms of entitlement. By taking the shilling, the citizen not only entitles the attribution of the commitment but also entitles designated officers to sanction him in the case of failing to serve. One consequence of mediating the practice of sanctioning with entitlement is that it brings the appropriateness of sanctions in question as a normative status, i.e. whether someone is entitled to sanction another. It is a further specification of the practice in terms of normative status.[24] A commitment is then not only a responsibility that is

---

[22] Ibid., 162. Brandom elaborates that this practice was justified as a way to confer commitment to illiterate citizens but in actuality it was a highly abused practice that involved disguised recruitment officers circling taverns where drunken citizens with empty pockets would take the queen's shilling.

[23] Ibid., 34–46. The relationship of sanctioning to deontic scorekeeping deserves brief attention because it is a crucial feature for understanding how normative practices work. Positive and negative sanctions provide one explanation of what it means to assess performances as correct or incorrect. Sanctions can consist in withholding rewards or distributing punishment for incorrect performances and the provision of rewards or withholding of punishment for correct performances. Brandom divides sanctions into two kinds: nonnormative or normative. The first kind involves responses to performances that ultimately depend on the disposition of the sanctioner. Imagine a household where the behavior of lying has been designated as wrong but no specific punishment has been specified. Thus, when a child lies the parent might be disposed to spank the child one day and yell at the child the next day. This kind of sanction seeks to negatively reinforce the disposition that produced the performance without yet normatively defining what sanction is appropriate. What makes this sanction nonnormative for Brandom is that it can be explained in purely naturalistic terms by an observer without reference to a normative specification. Normative sanctions consist of responses to assessments that are defined by further changes in normative status, i.e. withholding of subsequent entitlements or removing preexisting obligations, and thus must be defined in reference to the internal workings of the normative practice. When the child lies, he loses his or her entitlement to give reports that will be taken as true by his parents. Normative sanctions do not have as direct a relationship to the reinforcement of a disposition in the ways that nonnormative sanctions do: "In such cases one is rewarded or punished for what one does "in another world" –by a change in normative status rather than natural state.

[24] For Brandom, a practice becomes more sophisticated as it become further defined by normative statuses. One only has to think about how policies evolve within an organization to specify additional commitments and entitlements related to it.

appropriately attributed by others but it is also an undertaking that authorizes others to hold the undertaker responsible in an appropriate way.

In accounting for the source of deontic statuses, Brandom proposes that they are instituted by practical *attitudes* of taking and treating each other as committed or entitled. These attitudes are not arbitrary in a normative practice and yet their normativity is not explicit in the form of rules. Following the Wittgensteinian, pragmatist line of tracing rules back to practice, Brandom calls norms that are implicit in practices *proprieties* and these are what govern deontic attitudes. The two basic deontic attitudes are *attributing* and *undertaking* (the latter is sometimes referred to as *acknowledging*) commitments and entitlements. A person can *attribute* entitlements and commitments to others. For example, a genetic counselor implicitly attributes the commitment to patients that they believe genetic information is important to understand. A person can *undertake* a commitment by making a claim and offer entitlement by uttering reasons for the claim. A genetic counselor undertakes commitments, e.g. risk assessments, about the patient's fetus and then offers reasons for these commitments. Brandom understands attributing as more fundamental than undertaking because our ability to track ourselves begins with the ability to track others. What undertaking makes possible are authorized attributions of entitlements and commitments and also entitlements to sanction under certain circumstances. This pragmatic emphasis on attitudes, sometimes called methodological pragmatism, is one of the features of Brandom's work that I think can be especially helpful for reflections on genetic counseling. Its focus on practical attitudes governed by implicit norms provides a detailed vocabulary to analyze and evaluate the implicit norms that govern scorekeeping attitudes of those who participate in genetic counseling.

Deontic statuses and attitudes are the raw materials for explicating the practice of giving and asking for reasons. When an assertion or claim is made, a commitment is undertaken. Brandom calls this an assertional commitment. Assertional commitments entitle others to attribute that commitment as well as hold the asserter responsible for what the commitment entails. Unlike other kinds of commitments such as practical commitments, assertional commitments can be inherited by their attributers. This special feature will be addressed below. The broad context that allows the significance of assertional commitment to be worked out is the "game of giving and asking for reasons," a central concept that Brandom borrows from Wilfrid Sellars.[25] An assertion is made within this practice and plays the dual role of (1) being a reason (2) or standing in need of a reason. Claims or assertions are paradigmatically formatted as declarative sentences such as: *You have a 1/106 risk of giving birth to a child with Down syndrome.* Henceforth *P*, this sentence when uttered in a genetic counseling session presupposes and prompts the need for deontic, or mores specifically, conversational scorekeeping.

Conversational scorekeeping involves tracking the significance that utterances and non-linguistic behaviors have on the perspectives of those involved in a

---

[25] Wilfrid Sellars, "Some Reflections on Langauge Games," *Philosophy of Science* 21, no. 3 (1954): 204–28.

A Pragmatic Theory of Communication 59

conversation. Individuals inevitably have different commitment sets, otherwise there would be no reason to communicate. Keeping score in a conversation involves "keeping two sets of books."[26] When the genetic counselor utters P to Debbie, the genetic counselor has undertaken a commitment to P by asserting it. This entitles Debbie not only to attribute P to the genetic counselor but also to acknowledge P as a commitment she should undertake unless there is a reason to challenge P. Since the genetic counselor is a professional and Debbie a layperson, then Debbie will likely take the genetic counselor to be entitled to P by default although sometimes the accuracy of test results are questioned by patients. The question of what the consequences of P are for the HCP and the patient leads to the question of the content of P, its meanings.

What Is Meaning?

Brandom's answer to the question of meaning is inferential semantics. In the example of conversational scorekeeping above, the significance of P is understood semantically in terms of its inferential significance. As a potential reason[27] for further claims or as a reason in need of justification, *P's meaning* depends on the practice of inferring or, as Brandom puts it, "assertions are fundamentally fodder for inferences."[28] The pragmatic significance of a sentence, its meaning, is its inferential articulation in the practice of giving and asking for reasons. Inferences can refer to logically explicit inferences that are staples of introductory logics classes, especially in the form of conditionals, but more importantly for this project, inference refers to material inferences that can be observed in everyday exchanges and requires no logical vocabulary. Material inferences involve proprieties that govern the practical attitudes involved in linguistic practice. When the genetic counselor utters P, Debbie could have responded, "I'm in danger of having a sick baby" and this response is an inference from P. Whether it is a valid inference or one that should be challenged is an important question but it does not change its status as a kind of inferential doing. One of the responsibilities a genetic counselor has in offering P is to be fluent in the kinds of inferences that should and should not be made from P. If P is involved in a set of inferential relations, then these deserve further specification in terms of structure.

Brandom understands every sentence to be governed by an inferential network consisting of three kinds of relations: commitments, entitlements and incompatibilities. If I assert a sentence, then I am committed to, entitled to and prohibited from endorsing other claims whether I am aware of it or not. Let's take the proposition P.

---

[26] Brandom, 590.

[27] Brandom is not claiming that every sentence is uttered to provide a reason or every sentence ascribed should be challenged but only that they have the potential to be a reason for some further claim or action and the potential to be challenged.

[28] Brandom, 168.

To exhaustively survey these relations, one would have to identify all the premises that could commit, entitle and preclude entitlement to concluding P; and identify all the conclusions that must, can and cannot follow from P. No one within a given discursive community would have a complete grasp of all these relations, but professionals who offer sentences like P are certainly expected to have a significantly better grasp of them than lay persons. What limits or constrains the set of premises and conclusions to and from P are the contextual factors related to its actual claiming. Borrowing from Michael Dummett, Brandom broadly characterizes the meaning of a claim as the inferential articulation of the circumstances and consequences of its assertion. For example, P is easily recognized as a sentence appropriate to health care circumstances and more specifically to circumstances that involve pregnant women. What commits or entitles a HCP to utter these words is an important question as is the question of the consequences of uttering P from the HCP's view. Also of great importance are the circumstances and consequences of the patient that might serve as premises and conclusions to and from P once she is entitled to the claim. To say that a sentence stands in these relations is not yet to explain how one person inherits meanings from another person.

Brandom characterizes acquisition of inferential relations in which a sentence stands as the *inheritance* of deontic statuses. This process has an intrapersonal and interpersonal dimension. Let us first look at intrapersonal inheritance. If I wanted to justify P, I could work through the algorithms to arrive at P or attempt to grasp the full consequence of P for my preexisting commitments. In this dimension of inheritance, I possess these deontic statuses as a result of undertaking the commitment and actually doing the inferential work. The second dimension is interpersonal. When the genetic counselor offers P to Debbie, it introduces the interpersonal dimension of inheritance. The linguistic performance of asserting P has both pragmatic and inferential significance. First, it licenses the patient to attribute P to the counselor. Most often this licensing is implicit but it could be made explicit in the form of an ascription: 'The genetic counselor believes of me that I have a 1/106 risk of having a child with Down syndrome.' Since the claim is offered as true, the hearer of the claim is also being *licensed* to endorse it and incorporate the claim into her commitment set. How is the patient entitled to this claim? She inherits possession of its entitlement from the counselor who can presumably justify the claim. Most often, the patient infers that P has a default status of being true because the genetic counselor is assumed to be a reliable reporter. But the issue of whether P is true, whether the genetic counselor is entitled to it can and does arise. Some patients question the accuracy of genetic screens and tests.

The possibility of challenging P points to what Brandom calls the default-challenge structure within linguistic practice. The default component acknowledges that many statements we produce and consume have a default status of being true. Without the possibility of a default dimension, we would live in the world of the radical skeptic searching for and doubting the existence of truth as a property of assertions. The challenge aspect refers to the possibility of questioning someone's entitlement to a commitment. Without the possibility of challenging an assertion, there is no normative appraisal, no responsibility, no linguistic practice, no discursive communication.

The details of any given default-challenge structure depends on the context in which a discursive exchange occurs. What is said at a dinner party is not usually held to the same level of scrutiny as what is said at a public hearing. The default-challenge structure of genetic counseling has many features typical of a professional-client relationship. Given that professionals are expected to have expertise rooted in esoteric knowledge, clients often attribute a default status to a professional's statements. The difficulty in professional-client exchanges is knowing when to rely on what Brandom calls "person-based authority" and when to rely on "content-based authority." The former requires an act of deference and the latter requires an act of inference. The power imbalance in professional-client relationships is well established in the scholarship but less attention has been given to how this works in communication practices. How should we understand the movement between deferential activity and inferential activity? What will be shown below is that for the patient to understand and adapt to P in terms of the model presented here requires deferring and inferring. Before explicating how these alternating activities work, the relation between default-challenge structures and sanctioning must be elaborated.

A contiguous aspect of the default-challenge is the possibility of normative sanction. If the genetic counselor elicits incompatible claims from a patient, then the counselor might be less likely to attribute a default status to the patient's subsequent responses. For example, the patient may characterize her parents as healthy but in response to follow-up questions may reveal that her mother had breast cancer and her father has high-blood pressure. If more and more incompatible claims are detected, then the counselor may withdraw the presumed entitlement to make any claims that have a default-status. This across-the-board withdrawal of entitlement could correlate with a breach in the therapeutic relationship or questions about a patient's capacity to give consent. Patients also undertake the authority to challenge and sanction the HCP who undertakes genetic counseling. In a genetic counseling session that I observed, the physician recommended that the patient not undergo amniocentesis unless she would terminate the pregnancy. The patient challenged this recommendation claiming that the results from the test would reduce anxiety and allow her to sleep. In terms of the deontic model, the patient withdrew the physician's entitlement to the recommendation *in respect to her circumstances*. This withdrawal is a kind normative sanction that implicitly says: Whatever prior experiences entitled you to this recommendation, they do not entitle you to it in my case. In this discussion of default-challenge structures and sanctioning, the perspectival nature of meaning is suggested and deserves further explanation.

In interpersonal conversation, the inferential significance of a claim is relative to the background commitments of the producers and consumers of the claim. If claim P is undertaken by a genetic counselor, then it has inferential significance from the standpoint of both the genetic counselor and the patient. Let's take the perspective of the genetic counselor first. The genetic counselor's possession of entitlement to the claim's *accuracy* can be traced back to algorithmic calculations. By undertaking P the health care provider is automatically committed – think deductive commitments – to other commitments such as: (1) Your risk is lower than 50–50 but higher than the general populations risk of 1/733 of giving birth to a child with Down syndrome (2) You have a just under a 1 % risk of giving birth to a child with a chromosomal abnormality. In terms of entitlement-relations, the question is what does P *entitle* – instead of commit – the HCP to infer, an inference that is susceptible to challenge by countervailing evidence.[29] The reference to Down syndrome brings a host of inductive inferences about the clinical and social characteristics of a child with this condition. Although cognitive disabilities and low muscle tone in newborns with Down Syndrome occurs in close to 100 % of cases, they must still be classified as entitlements because rare circumstances do occur where these are not present.[30] A likely inference that a genetic counselor would offer is: "You have a 1/106 risk of having a child with mild to severe cognitive disabilities." Other entitlements would involve substituting in the underlined section clinical findings such as: 90 % of children with Down syndrome have significant hearing loss[31]; 30–40 % of children with Down syndrome have a congenital malformation in the heart or gastrointestinal track. The social characteristics of children with Down syndrome, e.g. prospects for independence,[32] effects on families,[33] are also entitlement-relations. Selecting and offering these consequences of P has come under increasing scrutiny by groups advocating for the disabled.[34] A major concern is that the primary criteria for interpreting Down Syndrome is to establish it as a medical condition that is justifiably the object of risk assessments. Finally, the incompatibilities of

---

[29] Brandom gives the example of a well made match and the entitlement to the inference that when struck against the appropriate surface it will light. He notes extremely cold circumstances in which such an inference could be challenged.

[30] J. R. Korenberg and others, "Down Syndrome Phenotypes: The Consequences of Chromosomal Imbalance," *Proc Natl Acad Sci U S A* 91, no. 11 (1994): 4998.

[31] D. S. Mazzoni, R. S. Ackley, and D. J. Nash, "Abnormal Pinna Type and Hearing Loss Correlations in Down's Syndrome," *J Intellect Disabil Res* 38 (Pt 6) (1994).

[32] For a longitudinal study comparing persons with Down syndrome to perform 'activities of daily living, see M. A. Maaskant and others, "Care Dependence and Activities of Daily Living in Relation to Ageing: Results of a Longitudinal Study," *J Intellect Disabil Res* 40 (Pt 6) (1996).

[33] The effect a child with Down syndrome has on a family is complex. For a population study that compares divorce rates between Down syndrome families, families with other birth defects and families with no known disabilities, see R. C. Urbano and R. M. Hodapp, "Divorce in Families of Children with Down Syndrome: A Population-Based Study," *Am J Ment Retard* 112, no. 4 (2007): 261–74.

[34] For a journalistic article discussing the conflict of interests between Down syndrome advocates and OB/GYNs, see Amy Harmon, "Prenatal Test Puts Down Syndrome in Hard Focus," *New York Times*, 9 May 2007, 1. Adrienne Asch has been a consistent and forceful critic of the prenatal testing. See A. Asch, "Disability Equality and Prenatal Testing: Contradictory or Compatible?," *Fla State Univ Law Rev* 30, no. 2 (2003): 315–42. There is also evidence that counselors are not having undue influence: van den Berg, Matthijs, Danielle RM Timmermans, Johanna H Kleinveld, Jacques Th M van Eijk, Dirk L Knol, Gerrit van der Wal, and John MG van Vugt. "Are Counsellors' Attitudes Influencing Pregnant Women's Attitudes and Decisions on Prenatal Screening?" *Prenatal diagnosis* 27, no. 6 (2007): 518–524. More recently, qualitative attention has revealed a lack of attention to disability, Ellyn Farrelly and others. "Genetic Counseling for Prenatal Testing: Where Is the Discussion About Disability?" *Journal of genetic counseling* 21, no. 6 (2012): 814–824.

P include that set of sentences that one is *not* entitled to undertake as a result of its being uttered. For example, the client is not entitled to the inference: "I *will* give birth to a child with Down syndrome." Because Debbie might infer from *P* that the fetus has been diagnosed with Down syndrome, genetic counselors often make this incompatibility relation explicit. One goal of the genetic counselor is to help the patient *inherit* a set of inferences from *P* that are valid from the standpoint of the role of a genetic counselor whose commitments are, for the most part supplied, by the biomedical tradition.

P's meaning extends beyond the counselor's commitment set to the perspective of the patient. This set of meanings is often referred to as subjective or personal to distinguish it from the objective meanings of *P*. Often the objective set of meanings is privileged when discussing what *P really* means but this leaves the status of Debbie's perspective in question. Indebted to the Gadamerian insight of *meaning pluralism*, Brandom's theory would classify Debbie's repertoire of commitments as a legitimate context in which *P* can be interpreted or specified. His position does not mean that all contexts are created equally. Some are more relevant than others. The key normative question is whether the HCP and Debbie think her perspective is important for understanding *P*. Whether and to what degree the genetic counselor should try to understand the genetic information from the patient's perspective are crucial questions that are answered differently by each of the genetic counseling models under consideration. I will address these differences below.

The deontic scorekeeping model is an idealized representation of the practice of undertaking a rudimentary conversation. The account proposes that both scorekeepers keep two sets of books or more explicitly track the conceptual content from both perspectives in effort to try to understand the content in view.[35] The reason for this demonstrates the depth of Brandom's pragmatism. In his view, grasping conceptual content requires that a scorekeeper be able to coordinate or negotiate inferences from different contexts of interpretation. This will be developed below in the discussion about the dialogical structure of grasping a concept.

Does Brandom's model mean that the genetic counselor does not understand *P* if she does not grasp it from the perspective of each patient to whom it is offered? It does follow that the counselor does not understand *P* if she exclusively interprets it from her own perspective, but such a unilateral relation to the content is highly unlikely. More likely is that the HCP habitually navigates between a biomedical perspective and an idealized patient perspective or a conglomerate of actual patient perspectives that she has encountered or gleaned from studies. If this latter case is operational, then the experience of uttering *P* to patient after patient may still seem to fit more along the lines of the technical vision's model of communication as transmission. The meaning of *P* is standardizable in an assembly line of perspectives because the idealized patient perspective is substituted for the actual perspectives of individual patients. The technical vision's understanding of meaning suppresses the role of the recipient's perspective because it privileges the sender's perspective and fails to see the perspectival structure of semantic phenomena.

---

[35] The model does not claim that both do it equally well.

Brandom rejects the notion that any perspective should have a permanent privileged status in the interpretation.[36] The circumstances of the utterance inform the selection of contexts that are important for interpretation. For a genetic counselor to claim that she understands P in *this circumstance* requires that she be able to make inferences from Debbie's perspective.

The patient has at least two interpretive tasks. One is to try to understand P from the counselor's perspective; the second is to try to understand it from her own perspective. For Debbie, the meaning of P is obtained by presuming that P is true, i.e. attributing a default status and sort the consequences of P as the genetic counselor explains them. She must first track *P's* inferential significance for the genetic counselor. Why is the HCP saying this to me? What does the HCP think the significance of this information is? If Debbie accepts P as true, then the question of entitlement is answered in her deference to the HCP. What remains is the task of identifying the inferential significance of P from Debbie's perspective. Her religious commitments provide an interesting case. Debbie believes that God's will would be present in several kinds of outcomes presumably because she believes God is active in all circumstances. Because P indicates a risk of possible outcomes, it *necessarily* involves from Debbie's perspective the issue of God's sovereignty and protection. From Debbie's perspective, she *should* interpret P in relation to God's will. To finally understand P is clearly relative to both Debbie's deferential and inferential activities. Thus, *understanding* is a kind of practical grasp that suggest dialogical structures.[37]

Having rehearsed several important components of Brandom's model, it is now possible to make the links between Brandom's understanding of dialogue and White's notion of dialogical counseling. Brandom sets out a distinction between dialogical relations and dialogical processes.[38] A dialogical relation holds between premises from two different sources or voices that share in the arrival at a common set of conclusions. As Brandom puts it, "In this sense each of them has its 'say.' For the collaboration of the commitments of the two as it were interlocutors consists in their relation to their joint inferential consequences."[39] To show in what sense

---

[36] Brandom endorse an I-Thou social model of linguistic practice rather than an I-we account. In the latter, the communities position on a matter is the truth of the matter ; in the former, the truth of the matter is a negotiated status between individuals who constitute a linguistic community (p.590): "Mutual understanding a communication depend on interlocutors' being able to keep two sets of books, to move back and forth between the point of view of the speaker and the audience, while keeping straight on which doxastic substitutional and expressive commitments are undertaken and which are attributed by the various parties. Conceptual contents, paradigmatically propositional contents can genuinely be shared, but their perspectival nature means that doing so is mastering the coordinated system of scorekeeping perspectives, not passing something nonperspectival from hand to hand (mouth to mouth).

[37] The possibility that P could affect every commitment is evidence of Brandom's endorsement of *semantic holism*, the position that meanings cannot be easily divided into analytic and synthetic distinctions.

[38] Robert Brandom, "Hermeneutic Practice and Theories of Meaning," *SATS - Nordic Journal of Philosophy* 5, no. 1 (2004): 23.

[39] Ibid.

premises share conclusions, Brandom gives the analogy of sharing in the way Fred and Ginger share a dance and not in the way that marching soldiers share a gait.[40] The presence of dialogical relations should be more pronounced in professional/client encounters like genetic counseling because there are pronounced perspectival differences that exist because of specialized training. The technical knowledge of the professional should combine with personal knowledge of client to arrive at a conclusion. For example, Debbie's statement about God's will and potential outcomes is a clear example that a dialogical relation has been established:

Genetic Counselor: Here are four possible outcomes…

+

Debbie: God's will is involved in this pregnancy and any outcome related to it.

=

God's will would be involved in a miscarriage from amniocentesis or the birth of child with Down Syndrome

Not all dialogical relations are this surprising and some can remain implicit in the simple substitution of words such as 'baby' for 'fetus'.[41] A genetic counselor refers to a *fetus* as having a risk and the patient refers to the *baby* as having a risk. This substitution is made explicitly dialogical in the following two ascriptions: (1)'You said that the fetus is at such and such a risk.' (2) 'You said about my baby that he or she is at such and such risk.' The second ascription contains an implicit dialogical relation that reflects input from both patient and counselor. For White, an optimal outcome of dialogical counseling would be a set of conclusions that reflects the *appropriate* dialogical relations. What makes such relations possible is, at least in part, the dialogical process of reaching them.

Brandom proposes a continuum of perspectival gaps that can be more or less bridged by dialogical processes. On one end are wide gaps – the kinds of gaps that hermeneutic theory usually addresses – that involve significant historical and cultural barriers to interpretation; on the other end are the everyday gaps that exist between individuals who are close in many ways but still travel this world along distinct spatiotemporal paths.[42] One could theoretically extend this continuum further by placing completely identical perspectives at one end and completely incommensurable perspective at the other. At both extremes, no dialogical relations or processes are possible.

The perspectives of HCPs and patients are neither incommensurable or identical and thus are amenable to dialogical *processes*. Dialogical processes include interpreting texts/utterances and more fundamentally grasping conceptual contents. In the case of texts or utterances, the interpreter plays a role in specifying the meaning

---

[40] Ibid., 24.

[41] For a more thorough discussion of this specific example, see L. de Crespigny, "Words Matter: Nomenclature and Communication in Perinatal Medicine," *Clin Perinatol* 30, no. 1 (2003): 17–25.

[42] Brandom, "Hermeneutic Practice and Theories of Meaning," 26.

of what some one else says or wrote. Two sources can be identified in the process of establishing the meaning of X. In the case of conceptual contents, a stronger and more global claim is being made:

> The most important notion of hermeneutic dialogue underwritten by inferentialist semantics is a different one, however. For according to the development of that view in *Making It Explicit*, [is that] practical grasp or understanding of conceptual content is the ability to *navigate* and *negotiate* between the different perspectives from which such a content can be interpreted (implicitly) or specified (explicitly)… When one can say both "S believes that a bunch of bloodthirsty fanatics occupied the village," and "S believes of a bunch of gallant freedom fighters that they occupied the village," one is calibrating claims (and concepts applied therein) according to the different doxastic perspectives of the author and the target of the ascriptions in a way that makes clear what inferential significance as premises they would have for each. Mapping different inferential significances, relative to distinct contexts, onto each other in this way is what taking them to be expressions of the same conceptual content consists in. For once again, it is the same conceptual content that is being attributed by the two ascriptions. Grasp of conceptual content in this sense is essentially dialogical, even in cases where one or more of the contexts in question is not associated with an interlocutor authorized to engage on its behalf in processes of expounding, expatiating, and answering for it.[43]

Dialogue on this view is constitutive of understanding and essential in the account of deontic scorekeeping. In Brandom's idiom, a practical grasp of genetic information in an actual genetic counseling session can only come from calibrating the claims of the HCP, the patient and possibly others in the room. The characterization of calibrating seems hypercognitive if one imagines a formal debate or literary scholar presenting a definitive reading but Brandom asks us to think in terms of material inferences and the perspectival structures built into the most basic exchanges. Thus, the example cited above of a patient substituting 'baby' for 'fetus' in reference to the same phenomena is a subtle but significant sign that a dialogical process is underway.

White's call for dialogical counseling appears more modest and not intended as global claim about what understanding entails but her social-perspectival account of autonomy prompts her call for dialogue and comports well with Brandom's position. Both view understanding and decision making as a social achievement. What Brandom provides for White and for the responsibility model is the elaboration of linguistic practice as an essentially normative, social, and dialogical phenomena.

Underwriting the Responsibility Model

Mary White's call for dialogical counseling promotes shared understanding and responsible decision making. She appropriates the work of H. Richard Niebuhr to show how a general account of selves as socially constituted can motivate a particular model of genetic counseling. What I have done above is to extend the theoretical

---

[43] Ibid., 24–25.

Underwriting the Responsibility Model 67

elaboration begun by White to both a larger story about communication and embodiment and to a more specific pragmatic and normative account of the structure of communication. Brandom's model depicts all acts of communication as interpretive events that have dialogical structures. The insights from these theoretical resources are sketched below in relation to the theses of the responsibility model. The model will be developed more in Chaps. 4 and 5.

1. *Goals: a) To help coordinate the meanings of genetic information with the patient b) To promote responsible decision making* – Genetic counselors should aim for a coordinated set of meanings about genetic information that reflect the appropriate perspectives. Below I will explicate the coordination of meanings in terms of the two tasks of navigating and negotiating perspectives. Achieving this first goal is similar to achieving the goal of an educated counselee as espoused in the teaching model but a few significant differences are worth mentioning. First, achieving a coordinated set of meanings requires a self-conscious awareness that that there are at least two legitimate perspectives trying to understand the genetic information. The genetic counselor should make inferences with P that cannot be made by the patient; the patient should make inferences with P that cannot be made by the HCP. Since the HCP is the professional with a technical grasp of the information, it might seem that only the patient needs to gain an understanding. The claim of the responsibility model is that the HCP does not adequately understand the genetic information *in this context* until some grasp of the patient's perspective is achieved. Although previous patient experiences contribute to the HCP's understanding, these have a provisional quality that can be revised by each new perspective. This view challenges the teaching model's picture of a genetic counselor transmitting objective information to the patient who then can ask questions or have misunderstandings corrected. Second, to obtain the outcome of coordinated meanings requires not only that dialogical relations are formed but also that an explicitly dialogical process is undertaken. The teaching model promotes dialogical relations in one sense. The patient receives the genetic information and then is free to commingle this input with his or her practical commitments to make rational decisions. This view aims to replicate the genetic information and the biomedical context in the patient who then can apply this information to their situation. The responsibility view aims to coordinate the meanings of the genetic information through explicit dialogical processes. An example will help illustrate the point.

When offering information like P to patients, a genetic counselor, who I observed, informed the patient that she was being given this information because her risk of having a child with abnormalities was greater than the risk of miscarriage from amniocentesis. The genetic counselor then asked the patient how she interpreted this risk. By making the biomedical perspective explicit and eliciting the perspective of the patient, this counseling maneuver provides the patient an opportunity to negotiate between perspectives. A negotiation between perspectives involves a consideration of meaning from two sources and determining which interpretation is worthy of semantic purchase. This example illustrates how dialogical processes can generate coordinated meanings.

Note that this example and this goal of the responsibility model do not aim to "understand the person" as Kessler proposes or to have an empathic understanding as comprehensive as Weil advocates. The genetic counselor attempts to understand the perspective of the patient in relation to the conceptual content and the constraints of the circumstances. Kessler and Weil could rejoin that they do not expect to achieve complete empathic understanding of the person and that this failure does not undermine the validity of the goal – one that has a family resemblance to the desire for shared interiors developed in Chap. 2. On the view endorsed here, a model should have achievable goals. Allowing for those exceptional cases where patients refuse to talk, the coordination of meanings is for the most part practically feasible.

Advocates of the psychotherapeutic model might also accuse the responsibility model of overemphasizing rational processes and neglecting emotional processes. Accepting the limitations of understanding does not mean that the emotional responses of the patient are ignored. If *P* causes a physiological reaction in Debbie, then that bodily response is potentially part of the meaning of *P*. Under Brandom's model, one can articulate the meaning of *P* and its relation to the physiological response of Debbie's body upon hearing *P*. It is not just the sounds of the words that caused the physiological response but the meaning of the words. What is the relation of the meaning of *P* to the physiological response? In this circumstance, *P caused* the physiological response. If this bodily response is called 'anxiety' or 'fear,' Debbie undertakes a *noninferential*[44] commitment that expresses (or reports) a bodily state caused by *P*. Henceforth, part of the meaning of P is related to the anxiety Debbie feels. Both Debbie and the HCP are responsible for coordinating what this anxiety means from their distinct standpoints. This description implies that an HCP who does not address the emotional states of a patient fails to take responsibility for an important consequence of offering P.

The second goal of the responsibility model is to facilitate responsible decision making. Two questions must be answered to understand this goal. Under what circumstances can it be said that a responsible decision has been made? To whom or what is the decision maker responsible? At the beginning of this chapter, Niebuhr's view of responsible action is used by White as a guide to promote responsible

---

[44] Perception is a term used by Brandom to refer to *noninferential* commitments expressed in observation reports. Noninferential commitments are triggered by what Brandom calls reliable differential responsive dispositions. On Brandom's view, we share these dispositions with land mines and thermostats but the latter do not undertake noninferential commitments. To explicate the important relation between noninferential commitments and inferential ones, Brandom compares a parrot that can be taught to say 'red' and the use of red by a competent language user. The parrot may be able to differentiate when to say, 'That is red' but has no understanding of the inferential significance of this move. When a competent language user claims, 'That is red', he or she knows the significance of the utterance, i.e. it is not a patch of green. If Debbie is implicitly or explicitly aware that she is anxious, then this attitude can be attributed as a noninferential commitment that has clear inferential significance for understanding *P*.

decision making in genetic counseling. These marks of responsible actions bear repeating: (1) every action is a response to a prior action (2) all actions –in contrast to behaviors – involve interpretation of what is happening, (3) an agent anticipates consequences of possible actions (4) fitting actions are acknowledgments of ongoing individual and collective narratives. The first mark requires that the genetic counselor be aware that he or she constitutes the prior action to which the patient must respond. The HCP is not a conduit through which objective information flows but rather the one who is endorsing the genetic information as true and useful and thus initiates the need for a response. Prior actions such as the referral also play a role but the genetic counselor is the most proximate agent to the patient's decision making. The second mark makes explicit the link between the two goals. Achieving responsible decision making depends on the ability to achieve an adequate coordination of meanings. The third mark requires that the genetic counselor help the patient think prospectively about the outcomes of her decision. In Debbie's case, the genetic counselor identified four possible outcomes for her to consider:

1. She can refuse the amniocentesis, avoid increased risk of miscarriage, and have a healthy baby.
2. She can refuse the amniocentesis, avoid increased risk of miscarriage, and have a special needs child.
3. She can undergo amniocentesis and miscarry a healthy baby.
4. She can undergo amniocentesis, an abnormality is found and then she must decide whether to continue the pregnancy.

Implicit in the presentation of these is the question of what it would be like to live with this particular outcome. Responsible decisions reflect accountability to future contexts. If the anticipated consequences are not compatible with the preferences and/or obligations that are anticipated to be operational in the future, then the action does not fit the context. And this brings us to the fourth and final mark. Responsible decisions are fitting or compatible with particular contexts. In the case of genetic counseling, two contexts are collaborating to make a decision, the biomedical context and the patient context. These contexts are suppliers of reasons for acting. Given the variation with patient contexts, are some decisions more responsible than others? What is not a responsible decision? A patient who flips a coin to decide whether to undergo amniocentesis is not on this account making a responsible decision even though one could make the argument that the outcome of the coin flip is a context. Whether the patient is entitled to his or her reasons for acting and whether the patient's perspective is sufficiently informed are important concerns when determining whether a decision is responsible. Who gets to decide whether a patient's decision is responsible?

This question is understood against the backdrop of the default-challenge structure that Brandom identifies above. In all three genetic counseling models, the patient's *pro-attitudes* and the beliefs they pair with them have a default status. A pro-attitude refers to those evaluative attitudes expressed as desires, preferences or obligations that constitute part of what forms an intention to act.[45] Giving a patient's pro-attitudes and commingled beliefs a default status is worked out differently by each of the models. The teaching model does this by educating or informing the client and then letting the patient make a decision without interference even if the decision may cause the patient harm. The psychotherapeutic model actively promotes the client's autonomy through empathic understanding and unconditional positive regard. Few circumstances warrant direct challenge. The responsibility model acknowledges the authority of the patient perspective by making it an integral part of the meaning of genetic information. At the same time its commitment to dialogue implies that other perspectives are needed to grasp the situation and to make a responsible decision. The responsibility model allows the genetic counselor more discretion to challenge the patient's decision making either implicitly or explicitly. Mary White elaborates this more interventionist stance in terms of dialogical counseling:

> In dialogical counseling the counselor is not a nondirective purveyor of genetic information; instead she acts as the interlocutor in dialogue, as the expert authority and sounding board for the individual faced with a genetic decision. Her role is not to minimize her involvement in decision making, but to promote a thorough and responsible deliberative process. As a partner in dialogue, the counselor provides the most current thinking on the medical, psychosocial, and moral aspects of reproductive decision making; helps her clients identify and clarify their priorities; dispels any misperceptions; and supports them as they make their decisions. If a client appears to be neglecting important factors or making a decision based on an unduly narrow perception of her alternatives, the counselor may introduce additional information and perspectives and the arguments for and against these views. If the counselor is concerned that a decision will be less than fully informed, is poorly reasoned, or may be regretted at a later date, she may encourage further reflection and discussion. If a decision is so at odds with prevailing social values that it could be considered ethically unacceptable, she may question her clients' reasoning and encourage them to think about the social consequences of the decision. Throughout the dialogue, the counselor does not debate, heckle, manipulate, or attempt to coerce a client's choice. Rather, her approach resembles that of a good teacher, therapist or friend, whose aim is to promote the flourishing of the individual in accordance with his or her unique personality and circumstances.[46]

The controversial features of White's description will be made more apparent in the next chapter when the value of nondirectiveness is directly addressed. Most of what White says above expresses what it means for the genetic counselor to facilitate responsible decision making. Her reference to teachers and therapists makes it clear that the responsibility model is not incompatible with those roles but the way these particular roles have been modeled in genetic counseling.

Two objections need to be addressed in regards to decision making in general. First, some patient responses are more like recognitions than decisions. Second, by

---

[45] See Donald Davidson, "Actions, Reasons, and Causes," *The Journal of Philosophy* 60, no. 23 (1963). Donald Davidson claims that action can only be understood in relation to a 'primary reason' defined as the pairing of a belief and a pro-attitude. When giving explanations for our actions (Why did you open the umbrella? …because it is raining), we often leave out the pro-attitude (I want to stay dry.).

[46] White, "Making Responsible Decisions. An Interpretive Ethic for Genetic Decisionmaking," 20.

focusing on the 'decision' the HCP may miss the most important concerns that need to be addressed. Churchill and Schenck refer to end-of-life circumstances where patients and their surrogate decision makers recognize without any observable inferences what must be done and their concerns tend not to be about medical decisions.[47] For example, family members sometimes spend no observable effort deliberating about when to remove life-sustaining treatment but give considerable attention to reconciling familial relations. One can imagine similar situations arising in prenatal genetic counseling where a pregnant woman sees or hears that she is carrying a severely deformed fetus. Whether and how the pregnancy will be terminated can become an issue at this juncture. For some women, the response to terminate does not come from an inferential process of decision making but rather as a recognition of their baby's condition. In such cases, issues of maternal grief are likely to be of more importance than a decision about termination. Avowing the responsibility model in this case would mean coordinating the meanings of this difficult finding leaving open whether or when a decision needed to be addressed.

One strength of the responsibility model is its adherence to a holistic structure of meaning. This stance requires the genetic counselor to be responsive to what genetic information and genetic anomalies mean for the patients in the broadest sense. It also requires an openness to interpretations that are different than one's own. For example, some women, for religious reasons, carry fetuses with severe and ultimately lethal anomalies to term despite the risk to their own health. Such cases demonstrate the ability of meanings to reach far beyond the ultrasound room where the anomalies are detected. Having provided a sketch of the two basic goals of the responsibility model, which will be developed in relation to Debbie's case in Chaps. 4 and 5, I turn to the assumptions necessary to justify these as obtainable goals.

2. *Based on assumption that clients come to share responsibility for understanding the genetic information and for decision making* – Although the psychotherapeutic model's perception that patients come for complex reasons is empirically accurate, it leaves implicit the kinds of commitments that must be attributed to the patient in order to begin the conversation. By leaving these assumptions implicit, the psychotherapeutic model prevents them from being explicated or scrutinized. One of the strengths of Hsia's account is that it acknowledges that people come for a variety of reasons and at the same time assumes that they should come to receive information. Likewise, the responsibility thesis acknowledges that genetic counselors have to make certain provisional assumptions about what motivates a patient to come for genetic counseling. Without such assumptions, the HCP would lack entitlement to engage the patient with the information.

Several challenges follow from assuming that a patient wants to share the responsibility of understanding and decision making. First, the counselor, in trying to coor-

---

[47] L. R. Churchill and D. Schenck, "One Cheer for Bioethics: Engaging the Moral Experiences of Patients and Practitioners Beyond the Big Decisions," *Camb Q Healthc Ethics* 14, no. 4 (2005): 393–98.

dinate meanings, may find that the patient is resistant to sharing her perspective. In these circumstances, the HCP should try to engage the counselee in different ways and allow for silence to play a role in the interaction. Sometimes patients have important reasons to be quiet. The need to be silent may indicate a need to focus energy on grieving or managing emotions.[48] If it becomes evident that the client does not want to communicate, then the counselor must disseminate the information and hope that the patient can take responsibility for it in non-observable ways. Peters' insight that dialogue can become tyrannical identifies the limits of dialogical processes and the need for an alternative stance. Second, the genetic counselor may experience patients who do not want help in decision making. Some may request privacy to make the decision. The genetic counselor must defer to the preferences of the patient if this occurs. Because the patient ultimately must live with the consequences of the decision, she has the authority to trade in explicit dialogical process for implicit or internal dialogical processes. The third challenge moves in the opposite direction. Some patients will want or expect the HCP to take all of the responsibility. This attitude is sometimes expressed in the 'what would you do' question or in the genuine surprise of being expected think about and respond to the information.[49] The genetic counselor should express in an affirming manner her expectation that the patient take responsibility for the situation. The HCP should make explicit her willingness to share in the process of decision making if the client will provide a place to start.

3. *The model assumes that the patient can participate in a dialogical process of grasping the genetic information and making responsible decisions.* – Whereas Kessler proposes that genetic counselors should have a complex set of psychological assumptions, the responsibility model makes assumptions about capacities to engage in communicative practices. Two key observational skills are needed to undertake this assumption. The first involves recognizing whether a patient is minimally capable of understanding and decision making in the context of the session. Although no clear cut litmus test is available, the session should start with highly accessible content, i.e. small talk about parking or the waiting room, to assess basic conversational skills. To be able to have a conversation at all goes a long way in determining whether a patient can participate in a more complicated, dialogical processes. On Brandom's view, to be able to converse requires the appropriate application of concepts. These applications demonstrate a base level ability to make judgments. After initial competency has been assessed, then ongoing judgments about the level of competence have to be made. Gauging levels of competence facilitates the appropriate navigation of perspectives. One of the impressive qualities of the genetic counselors I observed was their ability to calibrate their conversations to their patients' abilities. Some

---

[48] Patricia T. Kelly, *Dealing with Dilemma: A Manual for Genetic Counselors*, Heidelberg Science Library (New York: Springer-Verlag, 1977).

[49] This question does not necessarily mean that the patient wants to abdicate responsibility for decision making.

patients come to the session having acquired a beginner's fluency in the medical discourse usually through internet research; whereas other patients do not know why they have been referred or what a gene or chromosome is. This variation in levels does not challenge the basic assumption that a patient who can participate in dialogical process can understand and make decisions about the information.

Kessler objects to the teaching model's overemphasis on cognitive or rational processes. The same objection could be levied against these assumptions about practical capacities. The concern behind this objection is that the emotional or psychosocial dimension of the interaction will be neglected. Having responded to this criticism under the first thesis, it should be added that this objection reinforces the binary thinking that it seeks to challenge. It reinforces the bright line between objective and subjective meanings. Although subjective meanings are acknowledged in the aforementioned models – especially in the psychotherapeutic approach, they are often characterized as private meanings rather than alternative frameworks that have been appropriated by the patient to report her experience of the conceptual content. For example, the feeling of guilt in response to the news that one's child has a genetic condition is consistently interpreted as psychological, but it can also be characterized normatively as an inference from any number of communal narratives about parents being responsible for what happens to their children. Patients, just like HCPs, stand in relation to social contexts that provide institutional narratives, values, and rules that influence thought and action. Dialogical counseling is a way of coordinating this conglomeration of meanings.

4. *Counseling task is to facilitate (1) navigation and negotiation of the appropriate perspectives for understanding the genetic information and (2) practical reasoning about what action to take in reference to the relevant sources of responsibility* – If coordinating meaning is the goal, then the tasks necessary to achieve such an outcome involves navigating and negotiating perspectives that are trying to understand the same conceptual content. Although much has already been said about this dialogical process, the example of pronoun use allows us to further demonstrate the distinction between navigating and negotiating perspectives.

Anaphora is an esoteric term best understood by the familiar example of pronouns. Pronouns have an anaphoric structure in that what they pick out depends on an antecedent that varies in terms of specificity of reference.[50] Brandom calls this asymmetric recurrence. Any lexical grouping, i.e. sentences, quantifiers, demonstratives, can perform this function and as a group are called proforms. Anaphora contributes to the present model by expressing "one of the central mechanisms by which communication is secured across the interpersonal gap created by difference in doxastic commitments."[51] As Brandom points out, this is generally significant for all communicators because it allows person A in a dialogue to refer to something that person B mentioned earlier even though person A is not familiar with or does

---

[50] Brandom, *Making It Explicit: Reasoning, Representing, and Discursive Commitment.*
[51] Ibid., 486.

not know what exactly is being referenced. One can see in genetic counseling that a patient may use various proforms to refer to opaque concepts that the genetic counselor introduces. HCPs should track whether patient responses contain proforms that allow for fluent conversation but conceal a lack of understanding. The usefulness of anaphora can be demonstrated by working through the way 'you' is used in *P*.

An anaphoric analysis of *P* raises an interesting question about the use of 'you' in terms of its antecedent especially in languages that do not distinguish between formal and informal use of second-person pronouns. If in the initial interaction between the genetic counselor's and patient's names were exchanged, then the 'you' is most likely assumed to refer to the patient. However, an alternative specification could be that 'you' refers to persons whose markers when inputted into the algorithm produced the 1/106 probability. 'you' refers to one member of an aggregate that entitle the genetic counselor to assert *P*, and it could also reflect the attitude of a professional who sees many patients in one day. Such a usage would be susceptible to the criticism of treating patients as numbers but could only be interpreted as such in hindsight after much else had been said. The reading of 'you' as referring to the actual patient becomes more plausible if the health care professional actually tries to learn about the 'you' to which the probability is correlated. The normative uptake here is that the use of 'you' makes possible this equivocation of antecedents that allows the patient to think the counselor is referring to her specifically whereas the counselor could be referring to 'you' as a tokening of an algorithmic output.

Patients can also use pronouns to both navigate perspectives and at the same time conceal their ignorance. Take for example, "Does my baby have it?" would be an appropriate response to *P*, and the use of 'it' allows the patient to refer to Down syndrome without having to make any substantive substitutions. Compare this to the response, "I am in danger of having a sick baby" that substitutes 'sick' for Down syndrome. At minimum this substitution expresses that the vocal sound 'Down syndrome' means 'sick' to the patient whereas 'it' reveals almost nothing semantically. The anaphoric dimension of linguistic practice has received little attention in reflections on genetic counseling and, admittedly, is difficult to track in real time. It may prove more useful when doing transcript analysis for training purposes.

One reason for introducing anaphora is that it highlights a linguistic resource that allows genetic counselors and patients to navigate perspectives without yet negotiating them. One skill that a genetic counselor must have in identifying whether negotiations are occurring is to track whether the patient is making substantive *substitutions* for terms initially introduced by the HCP. Substitutional inferences are the proprieties that govern the substitution of singular terms and predicates. When sentences like *P* are uttered to the patient, he or she interprets it in part by substituting available terms that they consider to be equivalent. These substitutions allow the claim to then be interpreted in terms of their own background commitments. "I am (You have) in danger (a 1/106 risk) of having a sick baby (of having a child with Down syndrome)" is an inference from *P* that reflects substitutional commitments on the part of the patient. Restatements of this sort should signal to the health care provider the difference in perspectives and provide clues for follow-up statements.

The relation between substitution and meaning is significant for interview techniques that call for restating patient comments or sharing vocabularies.[52] Restating patient utterances should be understood normatively as part of the process of negotiating perspectives.

The notion that the HCP and patient perspectives can negotiate meanings may seem to threaten the objective status of the genetic information. This threat is often avoided by drawing a bright line between objective meaning and personal meaning, biomedical content and psychosocial content. The problem with drawing such a line is that it underestimates the need to negotiate or adjudicate the meanings of the biomedical and patient perspectives as they form dialogical relations in a specific context. For example, one consistent finding in patient responses to probabilities is that they interpret these numbers in unpredictable ways.[53] This conclusion does not threaten the objective status of the proposition, i.e., its standardized algorithmic derivation, but it might bring other biomedical hypotheses into question such as whether this form of information is *helpful* in patient decision making. It is this commingling of objectively derived and practically applied commitments that makes the need for coordinating meaning especially important in genetic counseling. If the patient is presented with options, this meaning making takes on the form of practical reasoning that ends with a decision.

Practical reasoning is a crucial part of responsible decision making. From the HCP's perspective, the primary reason that *P* is offered to someone like Debbie is for her to use it in her practical reasoning about whether to undergo amniocentesis. *P* will play a part, large or small, in the process of practical reasoning as it is related to desires for healthy children, responsibilities to children already born or to divine commands about the treatment of miraculous gifts. A dialogical *relation* will form between *P* and the desires, responsibilities, or obligations of the patient whether a dialogical *process* is actually undertaken. If the HCP is responsible for facilitating decision making, then a basic understanding of some styles of practical reasoning will allow for a more informed dialogical process.

Brandom elucidates three distinctive patterns of practical reasoning rooted in what he calls material inferences. Jeffrey Stout gives the following examples to illustrate Brandom's patterns:

(a) Going to the store is my only way to get milk for my cereal, so I shall go to the store.
(b) I am a lifeguard on the job, so I shall keep close watch over the swimmers under my protection.

---

[52] For examples of arguments for the importance of restating, see Baker and others, 55–74. and de Crespigny: 17–25.

[53] See Kessler and Levine, "Psychological Aspects of Genetic Counseling. Iv. The Subjective Assessment of Probability."; A. Lippman-Hand and Fraser, "Genetic Counseling – the Postcounseling Period: I. Parents' Perceptions of Uncertainty.", A. Lippman-Hand and F. C. Fraser, "Genetic Counseling: Parents' Responses to Uncertainty," *Birth Defects Orig Artic Ser* 15, no. 5C (1979).

(c) Ridiculing a child for his limp would humiliate him needlessly so I shall refrain from doing so.[54]

Inference (a) is representative of that type of practical reasoning that uses desire or preferences as a premise; (b) exemplifies practical reasoning founded on role-specific responsibilities; (c) involves an inference from an unconditional *ought* what Brandom calls an "agent- and status-blind pattern of endorsement of practical inferences as entitlement-preserving."[55] All three of these inferences are materially sound without being *logically* sound in a formal sense. Logic plays an expressive role that helps make explicit the premises of material inferences. Stout demonstrates:

> Suppose I added to (a) a statement expressing my desire to have milk for my cereal; to (b) the conditional that if I am a lifeguard, it is my *responsibility* to keep a close watch over the swimmers under my protection; or to (c) the principle that one ought not to humiliate people needlessly[56]

Most patients as do most people in general express themselves using material inferences without formal logic or making explicit revery relevant premise. One way an HCP can facilitate decision making is to make a patient's practical premises explicit especially when conflicts arises. When ethical problems arise, the source can be traced to a conflict *within* a particular pattern, i.e., conflicting desires, or *between* patterns, i.e., a conflict between a desire and an unconditional obligation. One tendency that should be avoided is to try to reduce practical reasoning to one style. If a patient is expressing her dilemma as a conflict between parental responsibilities, then the HCP should not ask 'What do *you* really want to do?' This question ignores the style of reasoning used by the patient. An example from Debbie's case illustrates the usefulness of Brandom's styles of practical reasoning.

Debbie indicates that she does not want to leave her present children with the responsibilities of care giving for a disabled child. She also claims that she does not want to place the baby at risk by undergoing amniocentesis. After the various outcomes have been identified, she introduces the concepts of miraculous gift and God's will. These concepts implicitly suggest responsibilities to a divine agent although these are not specified. The genetic counselor should observe that Debbie's role as a parent plays a large part in her practical reasoning. The HCP could point out to Debbie that she clearly takes her responsibilities as a parent seriously. A subsequent statement might inquire whether she feels equally strong about her responsibilities to her present children and to the fetus. Debbie's introduction of religious terms also warrants follow-up from the counselor. She is introducing another source of values in her role as a religious practitioner that she finds important in reasoning

---

[54] Jeffrey Stout, *Democracy and Tradition*, New Forum Books (Princeton, N.J.: Princeton University Press, 2004), 188. Stout's examples are based on Brandom, *Making It Explicit: Reasoning, Representing, and Discursive Commitment*, 245.

[55] Brandom, *Making It Explicit: Reasoning, Representing, and Discursive Commitment*, 252. Brandom acknowledges that moral philosophers tend to reduce moral reasoning down to one type of pattern. For example, Hume reduces moral reasoning to desire whereas Kant prefers to reduce moral reasoning to unconditional obligations.

[56] Stout, 188.

about the decision. One question that might be helpful is: 'Do you feel as though God will bless whatever decision you make?' This prompt might help Debbie clarify her statements about God's will and it also introduces the concept of blessing which may be implicit in Debbie's comments about God's will. Much more will be said in Chap. 5 about the peculiarities of addressing religion in genetic counseling. In this context, it is important to note religious belief as an important influence on styles of practical reasoning, one that is not exempt from the dialogical process in the facilitation of decision making.

5. *Relationship aims towards the mutual recognition of shared responsibilities* – In the commentary on the psychotherapeutic model, I suggested that mutuality might be defined as the mutual *recognition* of persons as distinct sites of authority. The concept of mutuality is further qualified here in terms of the mutual recognition of respective responsibilities. Both the genetic counselor and the patient should recognize their shared responsibility for understanding the genetic information and for decision making. If the genetic counseling relationship is mutual in the sharing of responsibilities, then the specification of these responsibilities shows differences and inequalities in perspectives. The patient will never understand the genetic information in the same way the genetic counselor does. This difference produces an epistemic inequality that gives the HCP a greater but limited authority over what it means to properly understand the genetic information. The genetic counselor will never have as much at stake in the decision making as the patient. This difference creates practical inequalities that give the patient a greater but limited authority in deciding what a responsible decision entails. These shared responsibilities and perspectival differences both define the relationship and generate the need for dialogical processes and relations or dialogical counseling.

Summary

The arguments of this chapter seek to develop an alternative model of genetic counseling that is an heir of the embodiment tradition of communication and receives further theoretical support from Brandom's theory of communication. Unlike the spiritualist tradition, the embodiment tradition does not aim for shared interiority. It accepts that we must navigate perspectival differences in order to coordinate a shared world. Brandom's model of deontic scorekeeping marks off the pragmatic middle ground between stances towards communication that embrace mutual understanding[57] or flee into radical otherness.[58] Here we are, different and in need, but with the resources to talk to one another. This attitude towards communication demarcates a space into which the responsibility model fits. Professional/client communication brings special challenges and responsibilities because of the marked differences in perspective. This feature is clearly present in genetic counseling. In the previous chapter, we reviewed the teaching and psychotherapeutic models that sought to overcome these differences through technical transmissions and empathic identification respectively. The responsibility model seeks to overcome them by coordinating meaning across different perspectives.

Having offered a new model and underwritten it with a theory of communication, the next two chapters will compare how all three models address two important issues in genetic counseling: nondirectiveness and spiritual assessment. Despite the lack of clear specification over the years and the growing discontent with it as a defining principle of genetic counseling, nondirective counseling continues to pervades the ethos of genetic counseling. By contrast spiritual assessment has emerged recently but often provides the HCP with vexing issues.

---

[57] For an exchange that makes explicit Habermas's emphasis on mutual understanding and Brandom's critique of Habermas's stance, see Jürgen Habermas, "From Kant to Hegel: On Robert Brandom's Pragmatic Philosophy of Language," *European Journal of Philosophy* 8, no. 3 (2000): 322–55. and Robert Brandom, "Facts, Norms, and Normative Facts: Reply to Habermas," *European Journal of Philosophy* 8, no. 3 (2000): 356–74.

[58] The relationship between Levinas's and Brandom's positions is not well understood and more research needs to be done in this area. Levinas's emphasis on inscrutable otherness and the role of embodiment, i.e. 'faces,' in founding the ethical relationship stands as a critique against an emphasis on social practices, i.e. linguistic practices, that bind us and allow us to identify with one another through language. I think Brandom and Peters partially avoid this critique in their resistance to cheap claims of mutual understanding.

Chapter 4
Genetic Counseling and Nondirectiveness

> We're supposed to ooze empathy, but stay aloof from decisions. Oh, I know I'm supposed to be value-free. But when you see a woman on welfare having a third baby with one more man who's not gonna support her, and the fetus has sickle cell anemia, it's hard not to steer her toward an abortion. What does she need this added problem for, I'm thinking'?
>
> So I try to put it in neutral, to go where she goes, to support her whatever her decision. But I know she knows I've got an opinion, and it's hard not to answer when she asks me what I'd do in her shoes. "I'm not pregnant," I say, "remember that."
>
> A social worker who trained me at Sloan-Kettering taught me something important: to clear my own agenda before I walk into the room, to let the patient set the agenda. It's the hardest lesson, and the most important one.
>
> - Rayna Rapp from interview with genetic counselor[1]

Nondirectiveness has been a core value in genetic counseling models for over 30 years[2] and yet there is little consensus about what it means. Nondirectiveness is a contested set of attitudes about the role a genetic counselor should play in helping a patient understand and make decisions about genetic information. Of its early advocates, Sheldon Reed – coiner of the term 'genetic counseling' – is the most influential. Although he never used the term 'nondirective,' he advocated the attitudes to which the term refers.[3] Fine cites this passage as an articulation of his early view that comports with nondirectiveness:

> They want to know what the chances are of another abnormality. We give them the figure if we have reliable one; otherwise we tell them we do not know the value. The parents often ask us directly whether they should have more children. This question is one we do not

---
[1] R. Rapp, "Chromosomes and Communication: The Discourse of Genetic Counseling," *Medical Anthropology Quarterly* 2, no. 2 (1988): 154.

[2] D. C. Wertz and J. C. Fletcher, "Attitudes of Genetic Counselors: A Multinational Survey," *Am J Hum Genet* 42, no. 4 (1988). For a historical overview related to eugenics, see R. G. Resta, "Eugenics and Nondirectiveness in Genetic Counseling," *J Genet Couns* 6, no. 2 (1997): 255–8.

[3] Beth Fine, "The Evolution of Nondirectiveness in Genetic Counseling and Implications of the Human Genome Project," in *Prescribing Our Future : Ethical Challenges in Genetic Counseling*, ed. D. M. Bartels, Bonnie LeRoy, and Arthur L. Caplan (New York: Aldine de Gruyter, 1993), 102–3.

answer because we cannot. The counselor has not experienced the emotional impact of their problem, nor is he intimately acquainted with their environment. We try to explain thoroughly what the genetic situation is but the decision must be personal one between the husband and wife, and theirs alone.[4]

Reed's view equates nondirective counseling to withholding advice about decision making. In an ongoing discussion, this articulation represents the dominant camp, one that is compatible with the teaching model of genetic counseling. Motivated by the psychotherapeutic model, Kessler along with others in the field articulate an alternative view of nondirectiveness. Seeking to define it in positive terms of what should be done, he elaborates the approach as a set of counseling skills that seek to promote the autonomy of the client. For example, genetic counselors should evaluate and point out the decision making strengths of clients. As these two understandings compete to articulate this constitutive value of the practice, recent challenges have proposed that nondirectiveness should no longer play such an important role.

John Weil's call for the post-nondirective era and the omission of nondirectiveness in the 2006 NSGC definition of genetic counseling represent individual and institutional efforts to reconceptualize the relation between nondirectiveness and genetic counseling. Weil acknowledges that competing definitions exist and that robust definitions along the lines of Kessler's view are new wine in old skins. He suggests that nondirectiveness as a defining term of genetic counseling is a "historic relic" that has outlived its usefulness and that its replacement should be to bring the psychosocial dimension into every phase of the counseling process.[5] The NSGC committee decided that nondirectiveness should be omitted in its 2006 definition for several reasons. An NSGC workshop in 2003 was held to discuss the relation between nondirectiveness and genetic counseling. Conclusions from the workshop included that its role must be clarified; that genetic counseling is a "psychoeducational interaction" incompatible with tenets of nondirectiveness; and finally that nondirectiveness must be sensitive to the growing numbers of clinical circumstances such as cancer genetics that warrant recommendations from the provider.[6] Omitting nondirectiveness in the 2006 definition is an acknowledgment of all of these factors but raises the question about the long-term consequences of moving away from this core value.

Many factors have contributed to the rise and possible fall of nondirectiveness as a defining principle of genetic counseling. At stake are issues that extend beyond narrow professional interests. As the U.S. history of nondirectiveness shows, the first genetic counselors adopted this stance to create separation from the mandated eugenic policies of the early twentieth century. In this chapter, I offer a brief history of nondirective genetic counseling to establish the normative trajectory of the

---

[4] Ibid., 103.

[5] J. Weil, "Psychosocial Genetic Counseling in the Post-Nondirective Era: A Point of View," *J Genet Couns* 12, no. 3 (2003).

[6] J. Weil and others, "The Relationship of Nondirectiveness to Genetic Counseling: Report of a Workshop at the 2003 Nsgc Annual Education Conference," *J Genet Couns* 15, no. 2 (2006).

conversation. I then elaborate and evaluate how the teaching, psychotherapeutic and responsibility models specify and justify nondirectiveness. I conclude that the responsibility model offers the most adequate view of the feasibility and normative importance of nondirectiveness.

A Brief History of Nondirectiveness

In the first 40 years of the twentieth century, a policy of preventing the 'feeble-minded' from breeding was promoted by many U.S. scientists and politicians and by many of their European counterparts.[7] This stance was justified by the concern that the "normal-minded" and elite were producing far fewer offspring than the "moron or high grade feeble-minded class."[8] These eugenic concerns and goals directed early conversations between those who had genetic knowledge and those who "needed" genetic knowledge. They also led to mandated policies of sterilization. The U.S. on many accounts was a leader in a eugenics movement whose international culmination can be found in the Nazi political initiatives towards Jews in the 1930s and 1940s.[9] As Kevles points out, "In Germany, where sterilization measures were partly inspired by the California (sterilization) law, the eugenics movement prompted the sterilization of several hundred thousand people and helped lead of course to the death camps."[10] Eugenic initiatives were based less on rigorous science and more on insufficient evidence that justified prejudicial policies against certain groups.

Against this trend, some in the scientific community, e.g. Lionel Penrose (British) and James V. Neel (American), established the medical value of genetics with rigorous scientific methods. These initiatives would eventually materialize as heredity clinics in the name of "preventive medicine" rather than eugenics. Some have pointed out that these clinics were still vulnerable to the charge of being eugenic in terms of having the negative eugenic goal of reducing the number of children with birth defects; at the same time, preventive medicine promoted voluntary screening and sterilization as well as "sympathetic counseling."[11] The move from state mandated to medically voluntary practices reflected the gradual learning of an international lesson about eugenics: under the guise of objective scientific knowledge, genetic discourse can be used to justify unthinkable atrocities in the political realm. The various political manifestations of eugenics in the first half of the twentieth

---

[7] Diane B. Paul, *Controlling Human Heredity, 1865 to the Present*, The Control of Nature (Atlantic Highlands, N.J.: Humanities Press, 1995), 70.

[8] Ibid., 62.

[9] Daniel J. Kevles and Leroy E. Hood, *The Code of Codes : Scientific and Social Issues in the Human Genome Project* (Cambridge, Mass.: Harvard University Press, 1992), 11. Kevles' book is one of numerous accounts that depict the U.S. as an early implementer of eugenic ideas.

[10] Ibid., 10–11.

[11] Fine, 102.

century remain a ghostly presence in all subsequent conversations and initiatives involving genetics.

In the 1950s, academic geneticists sought to sever ties with mandated eugenics in organizations such as the Dight Institute. They implemented voluntary screening and sterilization. In a 1952 article that presented several physicians discussing genetic counseling cases, Sheldon Reed weighed in on the issue of giving advice:

> If the genetic counselor is to advise the couple as to whether they should have more children or not, he must consider all the psychological, social, and economic factors he can find out about. It is an extremely rare case, which, in the experience of the Dight Institute, can be advised outright, as to whether or not reproduction is indicated. The counselor has never suffered the particular circumstances which the parents of the affected child suffered and therefore cannot completely understand their feelings.[12]

Earlier in the article Reed acknowledges the possibility of psychoanalysis by challenging its feasibility: "It is not practical to expect that extensive psychoanalysis will be possible in genetic counseling cases, but the general reactions of both parents to the situation can be obtained."[13] What this indicates is that Reed and other genetic counselors who pioneered the ethos of nondirectiveness were aware of the psychological needs of patients and the possibility and improbability of using rigorous psychotherapeutic approaches to genetic counseling. This new sensitivity to the client's perspective is an indication of the movement away from politically mandated eugenics even as these same counselors still endorsed the eugenic goal of preventing birth defects.[14] What had shifted was the locus of authority. More deference was being given to patient perspectives in the pursuit of "dysgenic goals."[15] Sorenson claims that most research geneticists and physicians during this period trusted that their patients, many of whom had experienced the challenges of having an affected child, would make rational decisions consistent with goals of preventive medicine.[16]

In 1969, the training of masters-level genetic counselors began at Sarah Lawrence College. As discussed in Chap. 2, in the formative years of this institution's curriculum a Rogerian approach to counseling was introduced that included nondirective counseling. In this context, genetic counseling appropriated a version of nondirectiveness that was informed by a substantive psychotherapeutic system. The masters-level genetic counseling modeled by Sarah Lawrence spread as did the influence of the Rogerian approach during a period where the need for prenatal genetic counseling was rapidly increasing.[17]

Two key factors precipitated the increase in numbers who received prenatal genetic counseling in the 1970s. First, diagnostic (amniocentesis) and screening

---

[12] Lee Dice, "Genetic Counseling," *Am J Hum Genet* 4, no. 4 (1952): 339.

[13] Ibid.

[14] Resta, "Eugenics and Nondirectiveness in Genetic Counseling," 256. Resta presents compelling quotations from several geneticists who promote nondirective counseling and eugenics aims.

[15] Fine, 103.

[16] Sorenson, "Genetic Counseling: Values That Have Mattered," 9.

[17] Marks, 18–22.

techniques (alpha feto-protein) improved during the 1960s. Second, the 1973 decision in Roe v. Wade that legalized abortion gave prospective parents new reproductive options. Genetic counseling guided by competing views of nondirectiveness – either from a teaching or psychotherapeutic perspective – was now being offered to families many of whom had no family history of genetic problems. During this period the foci of genetic counseling were primarily offering information about reproductive risks, fetal diagnoses and pregnancy options. The counseling demand would soon extend beyond the reproductive sphere as genetic tests were developed in other areas of medicine.[18]

The proliferation of genetic information and testing in areas outside prenatal care has challenged the centrality of nondirectiveness in genetic counseling. As Wylie Burke and others have demonstrated, nondirectiveness is traditionally considered appropriate in circumstances where the genetic test has high clinical validity but no effective treatment exists, e.g. Huntington's Disease (HD). Patients should decide whether such information would be valuable without treatment. In circumstances where the test has high clinical validity and effective treatment exists, e.g. PKU,[19] nondirectiveness has not been considered necessary. A third set of tests with low clinical validity and no effective treatment is generally not considered appropriate for clinical use by HCPs. A final set of genetic tests has uncertainty associated with the test's clinical validity and the effectiveness of treatment. It is this final set, which will continue to grow, that has raised important questions about the role of nondirectiveness in clinical encounters.

In the mid 1990s genetic tests became available to detect mutations in BRCA1/2 genes associated with familial forms of breast cancer. Unlike genetic tests for HD and karyotyping for chromosomal abnormalities, the BRCA1/2 tests did not predict with certainty that a woman would get breast cancer. Instead it provided a risk assessment or idenitfied a predisposition to develop disease. Unlike genetic testing of HD and karyotyping for chromosomal abnormalities, BRCA1 testing results come with treatment options that vary in terms of effectiveness. Consider, for example, the role of prophylactic mastectomy as a preventive measure.[20] When a woman is found to be a carrier of BRCA1, she has several options including increased surveillance and prophylactic mastectomy. Some have questioned the latter maintaining that the benefits of prophylactic mastectomy are overestimated and the potential harms underestimated.[21] Nonetheless, evidence is growing that some benefits are

---

[18] For the most comprehensive account of this history, see Stern, A. *Telling Genes : The Story of Genetic Counseling in America*. Baltimore: Johns Hopkins University Press, 2012.

[19] PKU stands for phenylketonuria and is a metabolic disorder where the body is unable to metabolize the amino acid, phenylalanine. The result is that this amino acid builds up in the body causing damage to the central nervous system. This build up and the severe cognitive disabilities that accompany it can be avoided by putting the infant on a life-long diet that is free of phenylalanine.

[20] A prophylactic mastectomy involves the removal of one or two breasts in order to prevent the occurrence of breast cancer.

[21] F. Eisinger, "Prophylactic Mastectomy: Ethical Issues," *Br Med Bull* 81–82 (2007): 7–19.

conferred by the procedure.[22] What status does nondirectiveness have in these circumstances? Is its primary role in the pretest conversations? Or should it be applied in eliciting preferences between potential treatments? At this juncture, it is important to see that genetic counseling models have had to adapt to the changing circumstances of genetic testing.

The rise of evidence-based medicine in the early 1990s has spawned investigations into genetic counseling asking whether nondirectiveness is a value that is actually being upheld in practice. Benkendorf has asked whether genetic counselors use indirect communication as a tactic to achieve nominally "nondirective" communication. Indirect communication involves substituting a direct expression e.g. the preference 'I prefer the window down' for an indirect one, e.g. 'It sure is cold in here' that requires the recipient of the communication to make the inference, 'You want the window down.' The second person does actually make this inference but the first person is clearly directing them there. The study found that genetic counselors do use indirect communication as a technique that permits the client to make her own inference despite the use of indirect statements as inferential vectors.[23] Probably the most important empirical study of nondirective counseling, published in 1997, was undertaken by Michie et al.[24] Acknowledging that no accepted definition of nondirectiveness exists, they identified directiveness in statements where the counselor gave directions or advice to a counselee and sessions where selective information giving or endorsement by the counselor was clear. Using this criteria, they asked both counselors and counselees to rate the genetic counseling session. The researchers concluded that nondirectiveness was not being uniformly undertaken and that lower socioeconomic clients as well as highly concerned clients were more likely to be directed in their decision making. Kessler responded to this study by stating that Michie's criteria for directiveness cast too large a net. For example, advice giving in genetic counseling was not always directive. The results of these studies in the 1990s and early twenty-first century have intensified the discussion about what nondirective counseling entails and its feasibility.

Whether nondirectiveness is achievable or definable remains an important question to which the aforementioned proposals by Weil and the NSGC's definition offer answers. The normative and empirical questions about this concept suggest either that it is losing or never fully had a grip on the profession or that it no longer needs to have primary role in defining this professional task. Will these proposals and the changing contexts marginalize the role of nondirective counseling or does it have a default status that will allow it to persist?

In the next section I elaborate how the three models under investigation specify nondirectiveness and address whether they justify or jettison its role in the changing

---

[22] T. R. Rebbeck and others, "Bilateral Prophylactic Mastectomy Reduces Breast Cancer Risk in Brca1 and Brca2 Mutation Carriers: The Prose Study Group," *J Clin Oncol* 22, no. 6 (2004): 1055–62.

[23] Benkendorf and others: 199–207.

[24] S. Michie and others, "Nondirectiveness in Genetic Counseling: An Empirical Study," *Am J Hum Genet* 60, no. 1 (1997): 40–47.

context of genetic counseling. Part of this elaboration involves showing how these approaches would evaluate the genetic counselor's performance in Debbie's case. I begin with the teaching model.

Nondirectiveness and the Teaching Model

In Chap. 2, I noted Sorenson's claim that the attitudes of the early academic geneticists influenced the principle of nondirectiveness in ways that remain operational in genetic counseling. As academics, they valued what Sorenson calls 'technical neutrality' a stance that incorporates the concern for standardization in scientific method and factual presentation in the pedagogy of the sciences. Sheldon Reed developed the initial connection between nondirectiveness and the teaching model, characterizing the genetic counselor primarily as an educator with limited knowledge to make recommendations. A genetic educator presumably with some expertise has a grasp and should carefully explain objective information such as risk assessments but cannot grasp or explain the perspective of the patient in the same way. What is implicit in Reed's view is also representative of the teaching view: HCPs who undertake genetic counseling are entitled to offer and explain genetic information to patients because of its objectivity but they are not authorized to offer judgments about the patient's circumstances because it involves subjective information to which they have little or no access. Genetic information is a set of objective messages that any geneticist would offer to any individual with the same genetic circumstances. It is in this respect 'neutral' towards the actual person receiving it. In the teaching model, the authority of the professional is inherited from the authority of the true and neutral information they offer. Sorenson suggests that this understanding of authority and neutrality was consistent with the academic dispositions of geneticists who pioneered the task of genetic counseling. Sorenson also suggests that physicians in contrast to scholars/research physicians value neutrality less and intervention more.[25]

One of the criticisms of the teaching model's interpretation of the nondirective value is that it only attends to what genetic counselors should not do. Before looking at these criticisms, it needs to be made explicit what exactly the HCP providing genetic counseling should not do and why these prohibitions are warranted by an educational emphasis. In Chap. 2, I introduced Hsia's account of information provision as a substantive paradigm for the teaching model.[26]

Hsia's account of nondirectiveness illustrates a negative elaboration of nondirectiveness. His work is helpful not only because it is a substantive presentation of the

---

[25] Sorenson, "Genetic Counseling: Values That Have Mattered," 10.

[26] Hsia's article represents the most articulate account of the teaching model of genetic counseling but he does not have much competition. Sorenson is another influential supporter of the "genetic education" model rather than "genetic counseling" model. I hypothesize that the lack of theoretical resources to elaborate the teaching model indicates its dominance.

teaching model but also because it reflects the perspective of an academic geneticist and pediatrician who does genetic counseling. This combination of training generates a stance that takes seriously the temptation to intervene in patient decisions and at the same time articulates practical reasons to keep a certain distance.[27] He issues a litany of 'nons' that nicely encapsulates his position: "I advocate the responsibility of genetic counselors to be nondirective, nonpsychoanalytic and nonjudgmental."[28]

In a subsection titled "What to Avoid," Hsia presents several sets of reasons for resisting the temptation to direct families in their decisions. The first set of reasons are motivated by the concern to maintain the respect or trust of the patient. Such concerns are related to issues of demarcating professional authority. If the genetic counselor gives a recommendation, then she is vulnerable to loss of respect in two ways. First, if the counselee does not take the counselor's advice and finds the other course of action acceptable, then she will lose some respect for the counselor's judgment. In the second and more consequential scenario, the patient takes the advice, e.g. forgoes amniocentesis, and finds living with the consequences unacceptable, i.e. caring for a child with Down syndrome. In this case, not only has respect for the counselor been lost but also the counselee must live with the long-term responsibilities and challenges of a decision that the professional authority recommended. These reasons for not offering advice to a family are based on the negative consequences of taking such directive action, but Hsia also acknowledges that the complexity of the circumstances raises the question of who should be authorized to make the decision regardless of outcome: "who *does* know the best course for a family to take?"[29]

A core premise in the teaching model's conception of nondirectiveness is that the genetic counselor is not in a position to know the best interests of the client.[30] Without explicit mention of the respect for autonomy, Hsia argues that counselees need to come to their own conclusions about the best course of action. He asserts that a genetic counselor cannot make a judgment about whether a counselee's decision is rational especially in decisions involving reproduction:

> The procreative drive of counselees, however, its intensity, constancy and durability are beyond the professional ken of the geneticist. Other aspects of the personal vagaries of counselees are equally imponderable such as the willingness or ability of a person to accept and adapt to a compromise solution and the strength of character needed for a person to cope with stressful life events." (181)

No matter how determined a genetic counselor is, she will never bridge this perspectival distance and thus lacks essential premises to make practical decisions on the patient's behalf. Hsia's skepticism about a counselor's ability to understand

---

[27] Hsia's account predates cancer genetics and thus is organized primarily around a set of counseling circumstances that involve reproductive decisions.

[28] Hsia, 169.

[29] Ibid., 181.

[30] Sorenson, "Genetic Counseling: Values That Have Mattered," 11–12.

patients' emotions, drives and character is consistent with his cautious attitude toward psychotherapy in genetic counseling.[31]

The short term nature of genetic counseling sessions and the complexity of undertaking psychological assessments have led advocates of the teaching model to challenge the use of counseling techniques in the genetic counseling. In "valuing nondirectiveness over advice giving and education over counseling," Sorenson makes explicit the connection between teaching and nondirectiveness and its tension with the psychotherapeutic model.[32] Hsia justifies these preferences by pointing to the potential harm a counselor could inflict without the proper training and experience in using counseling skills. Genetic counselors should have enough psychological knowledge to make a referral to a trained professional.

The third 'non' in Hsia's triad is being nonjudgmental. If a counselor should avoid delving into a counselee's decision making or psychological states, then it follows that they do not possess entitlement to morally judge the final decision. Subtle or overt challenges to a counselee's decision are rarely an appropriate move for a genetic counselor to make. In cases where the best interest of the counselee seems to be in jeopardy, e.g. seeking "quasi-fraudulent care," the counselor should still refrain from challenge.[33]

Hsia's distinctions between being nondirective, nonjudgmental, and nonpsychoanalytic are often conflated in references to nondirectiveness.[34] One of the contributions of his account is that it raises the question of how to demarcate the complex attitudes of the counselor. For example, these distinctions suggest the possibility of being nondirective and doing psychotherapy if the training and time were sufficient. In other words, even as psychoanalysis is dismissed in this model, the distinctions themselves raise the possibility that they are not incompatible as will be apparent below. Despite these benefits, most who refer to nondirectiveness from a non-psychotherapeutic standpoint, continue to lump these together. In terms of the teaching model, nondirectiveness is a term for many in the profession that licenses an avoidance of patients' emotions, ambivalence and decision making in relation to genetic information.

In the rest of this section, I want argue that Hsia's and Sorenson's positions implicitly rely on specific interpretations of pedagogy, autonomy, and meaning. Pedagogical neutrality, a traditional and highly criticized[35] stance within education

---

[31] Ibid.

[32] Ibid., 12.

[33] Hsia, 184.

[34] Bosk, 28–29. Bosk like Hsia thinks the distinction between nondirectiveness and value neutrality should be distinguished because physicians are rarely the former but certainly aim to adopt the latter stance.

[35] Henry Giroux, an influential scholar in critical pedagogy, claims in his article "Schooling and the Culture of Positivism: Notes on the History" that modern pedagogy has been dominated by positivism of which neutrality is a component. For a compilation of his works including the aforementioned article, see Henry A. Giroux, *Pedagogy and the Politics of Hope : Theory, Culture, and Schooling : A Critical Reader*, The Edge, Critical Studies in Educational Theory (Boulder, Colo.:

circles, is an assumption of the teaching model of genetic counseling.[36] As I tried to develop in Chap. 2, the teaching model appropriates a technical vision of communication where objective messages are sent through a noiseless channel in order to justify a certain pedagogical stance. One premise of this account is that schools and by extension educational events are neutral sites where universal facts about states of affairs are distributed. If this is true, then the best circumstances for learning require that the teacher bracket all his or her values and provide only neutral descriptions of what is the case. These efforts at perspectival constraint allow students to make inferences with defined concepts and stable facts in new situations. The idea of pedagogical neutrality addresses a central concern about teaching's relationship to autonomy.

Autonomy according to the teaching model is an individual's capacity to make rational decisions using objective genetic information in relation to her own commitments. Providing the circumstances for a person to exercise her autonomy is the explicit[37] motivation for pedagogical neutrality and the teaching model of genetic counseling. Autonomy on this account requires an individual to have available to him or her the terms, definitions, and facts to choose and use to make new inferences. The teacher can provide these resources but not the inferential activity that allows an autonomous person to actually use them. The teacher should provide all the resources for making new inferences, but she should not make the inferences for the student. The genetic counselor as teacher undertakes nondirective counseling when she presents the facts, figure and options in a way that does not advocate for a certain outcome. In short, she has bracketed her values when presenting the information and maintains a neutral stance as the student-client makes her own decision.

This notion of autonomy emphasizes not only an atomistic conception of the individual legislating his or her own conclusions but also an atomistic conception of meaning that funds the genetic educator's status of objective information giver. First, the set of meanings that is genetic information is presumed to be easily demarcated or self-evident. This picture of meanings suggests an image of handing the client an easily identifiable bag of marble-like meanings, each marble is self-contained, and the set is an accumulation of individual meanings. This premise suppresses the normative notion developed in the previous chapter that offering genetic information involves undertaking a commitment to offer P to the counselee and that the issue of the doxastic or practical entitlement to the assertion can be raised. These get articulated in actual patient questions such as: "How accurate is

---

WestviewPress, 1997). For an equally important figure in the area, see Paulo Freire, *Pedagogy of the Oppressed* ([New York]: Herder and Herder, 1970).

[36] Sorenson, "Genetic Counseling: Values That Have Mattered."; Hsia. Both Sorenson and Hsia assume that pedagogical neutrality is not only possible but the an important aim in genetic counseling.

[37] One of the main critiques of the relation between pedagogical neutrality and autonomy is its implicit motivation to maintain social control by offering the fiction of disinterested cites of authority, i.e. objective states of affairs.

the test/screen?" or "Why are you telling me this?" The second premise at work here is that the patient can add these new marbles to his or her own bag of marbles – a subjective bag that cannot be understood by the HCP – and then make a rational decision. This semantic view fails to take account both the holistic effect sentences can have on persons' commitment sets and the addition of a new commitments like, "I feel a pit in my stomach after hearing P." To avoid the responsibilities for these semantic and emotional effects on a client, the teaching model offers an account of nondirectiveness that depends on pedagogical neutrality and atomistic notions of autonomy and meaning.

The teaching model and nondirective counseling have an authoritative status that comes from playing a formative part in the development of genetic counseling; whereas the psychotherapeutic model and its understanding of nondirective counseling, which appeared slightly later in the history of genetic counseling, have played the role of challenger to the tradition. As a result, the psychotherapeutic model has needed to define itself over against the teaching model.

Nondirectiveness and the Psychotherapeutic Model

Nondirectiveness claims its conceptual home in psychotherapy, but its professional address is primarily located in the teaching model. This leads to the unexpected result that most of the criticism of nondirective counseling – as defined by the teaching model – comes from practitioners who endorse a psychotherapeutic or psychosocial model. Although Weil characterizes nondirectiveness as a historic relic, he also acknowledges that its meaning is indeterminate. He cites two dominant perspectives and aligns his position with the second camp. He characterizes the first set of meanings as centered around withholding advice and the second as promoting autonomy. He notes that both have conceptual debts to Carl Roger's development of the concept. In this section, I elaborate the way that nondirective counseling has been articulated by those who advocate it as promoting autonomy within a psychotherapeutic framework.

Carl Rogers introduces nondirective counseling into psychotherapy as an alternative to existing psychotherapeutic models that assessed, diagnosed and treated patients in a way that imitated biomedical models. The medical process he avoids proceeds like this: The patient is asked a series of questions by the physician. The patient's answers to these questions are used to make technical inferences that result in a diagnosis and recommendation for treatment. Rogers concludes that the psychotherapist's aim is to help a person become autonomous and to achieve congruence through healthy internal and external communication, a set of concepts discussed in Chap. 2. The biomedical model is not appropriate to achieve this goal, and in its stead he develops a more fitting approach called client-centered therapy. At its core is the counseling relationship. Nondirectiveness is a general strategy within this Rogerian framework.

This conception of nondirectiveness prescribes that the counselor does not direct the conversation in the counseling session. The counselee instead sets the agenda for the session and takes a lead role in the conversation. The counselor's role is to embody certain attitudes as he engages in conversation: unconditional positive regard, empathy and genuineness. If a counselor can genuinely achieve empathic identification with the counselee, bracketing all evaluations, then the counselee has an opportunity to observe and respond to his own perspective as it is empathically understood by the therapist. Through this method, a person is provided an opportunity to find her true self unencumbered by dogmatic and ideological conceptions of the self.

Similar to Rogers' assessment of the psychotherapeutic context, genetic counselors who advocate a psychotherapeutic model of nondirectiveness observe that genetic counseling does not fit neatly into the traditional biomedical model of diagnose and treat. Specifically, the offering of genetic information is not always followed by a recommendation for intervention. There are other similarities that make nondirectiveness appropriate. Receiving genetic information can cause deeply felt emotional responses and can disrupt attitudes about where one stands in the world. Often it requires a decision for a course of action e.g., termination of pregnancy or prophylactic mastectomy, decisions that involve intensely intimate spheres of human concern.

At the same time, genetic counseling has features that necessarily replicate the biomedical model and resist the Rogerian view of nondirectiveness. A genetic counselor often does ask questions and makes inferences or reports test results that indicate risk assessments or diagnoses. In this way, the agenda is significantly defined by the HCP, and the patient needs this direction. Whether this likeness makes genetic counseling incompatible with the Rogerian vision is a central question that the psychotherapeutic model must answer. Another key obstacle that must be overcome is that genetic counseling relationships are often one or two sessions whereas psychotherapeutic relationships in the client-centered model presumably develop over longer durations. The question raised by this difference is whether counseling skills such as empathic identification can actually be undertaken in such a short period.

Kessler has given nondirectiveness in genetic counseling the most careful reconsideration from the standpoint of the psychotherapeutic model and unlike Weil argues that it is the inevitable core of the profession. Its central focus is the promotion of the counselee's autonomy. Kessler acknowledges the debt that must be paid to Rogers and at the same time is willing to revise nondirectiveness to the degree that Rogers might not recognize it. I think it is accurate to describe Kessler's stance as a gradual movement away from the Rogerian vision toward a psychologically informed, technique-driven approach.[38]

Kessler addresses criticisms of nondirectiveness by providing a narrower definition of directiveness and a broader definition of nondirectiveness. His specification

---

[38] In his early writings in the mid-to later 1970s, Kessler is trying to put psychology on the radar screen of genetic counselors; in the 1990's, he is trying to nuance his position to make it more practical in relation to the contextual constraints of a genetic counseling session.

of directiveness is in part a reaction to Michie's study that concludes that a gap exists between the normative and empirical undertaking of nondirectiveness. Kessler suggests that directiveness is a special case of persuasive communication called "persuasive coercion." Persuasive communication entails the kind of influences we experience everyday that we do not consider as a threat to our autonomy; persuasive coercion – otherwise known as brainwashing or mind control – involves one or more components "1) deception 2) threat 3) coercion."[39] For example, if a genetic counselor intentionally withheld information about the risks of amniocentesis in order to encourage the uptake of the procedure, then the HCP would be deceiving the patient who might infer that amniocentesis has no risks. Kessler proposes that these elements of persuasive coercion should be the criteria for identifying directive actions in genetic counseling.[40] An important consequence of this restriction is that advice giving in certain circumstances should not be considered directive. Michie's findings that genetic counselors did not uniformly adhere to the value of nondirectiveness are less compelling if one accepts Kessler's restrictive view of directiveness because she used advice giving as a key code in identifying directive action. Another consequence of Kessler's stance is that more circumstances fall between directive and nondirective action. He acknowledges that much of what happens in genetic counseling does not neatly fit into these opposing categories.[41] Having narrowed the application of directiveness, Kessler seeks to restore the meaning of nondirectiveness.

Kessler offers this definition: "ND describes procedures aimed at promoting the autonomy and self-directedness of the client."[42] Kessler elucidates the tangled history that this term has with the Rogerian vision. In genetic counseling's appropriation of Rogers' term, Kessler admits much is lost. For example, counselees do not and should not direct the agenda in genetic counseling. This dimension of the Rogerian vision is lost, but the ultimate goal of promoting autonomy is retained. Kessler uses scenarios to highlight more specific goals and counseling skills required of nondirective counseling.

Nondirectiveness is an active form of genetic counseling that seeks to "raise their (the patients') self-esteem and leave them with greater control over lives."[43] What Kessler sees as his contribution to the conversation about nondirectiveness is that he reintroduces the competencies that counselors need to attain nondirective goals. He suggests several tactics for employing the strategy. The HCP must pay attention to competencies and accomplishments of patients and acknowledge these strengths verbally. The genetic counselor should encourage the client to talk more and reward clients for efforts toward autonomy. Passive notions that being a 'good listener' or being supportive are nondirective behaviors is wrong on Kessler's view because they do not actively promote autonomy. A counselor must help the counselee think

---

[39] {Kessler, 1997 #28}165.
[40] Kessler, "Psychological Aspects of Genetic Counseling. Xi. Nondirectiveness Revisited," 165.
[41] Ibid.
[42] Ibid., 166.
[43] Ibid., 169.

through decisions. During this whole process, the HCP must keep his or her own feelings in check and must keep the focus of the session on the patient.

In the rest of this section I want to show how Kessler's view of nondirectiveness retains vestiges of Rogers' program even as it jettisons much of the psychotherapeutic framework. Despite Kessler's disavowal of much of Roger's program, his most recent accounts of nondirectiveness still assume key commitments such as *unconditional positive regard* and *empathic listening*. Kessler proposes that nondirectiveness entails counseling skills such as (1) focusing on and understanding the client's perspective (2) assessing and articulating clients' strengths in decision making. One can imagine a reasonably large set of counseling skills from which Kessler is making his selection. His choices serve not only the goal of promoting autonomy but also reflect the premises that a client's autonomy needs validation and that his or her perspective can be assessed relatively quickly. The train of logic can be located within the stream of Rogerian thought that clients need the acceptance or approval of the counselor and that nonjudgmental understanding of the client's perspective can be gained empathically. The evolution of Kessler's thought moves in a direction of making these psychological stances implicit and making the counseling skills they serve explicit.

This move raises the question of whether counseling skills have to be drawn from a psychological framework. Without explicit psychological underpinnings, Kessler's counseling skills function in a similar way to the responsible communication skills I am suggesting. The difference is that I locate the responsibility model within an explicit normative theory of communication whereas it is not clear whether Kessler continues to avow the psychological accounts of communication of Rogers and others.

Nondirectiveness and the Responsibility Model

In Chap. 3, I offered the responsibility model as an elaboration and extension of Mary White's articulation of responsible decision making in genetic counseling. She gives several criticisms of what can now be seen as the teaching model's version of nondirectiveness. The thrust of her critique is that current understandings of nondirectiveness focus exclusively on protecting the client from the external constraints such as counselor coercion or interference from other agents. Genetic counselors understand themselves as upholding clients' negative right to an unfettered decision. The shortcoming of this model of nondirectiveness is that it neglects internal influences on the client's autonomous decision making such as emotional instability and implicit misconceptions. The corrective she offers is to make nondirectiveness a positive right that entitles a client and obligates a counselor to achieve the most informed and responsible decision possible. Her recommendation leads to a third specification of the nondirectiveness in terms of social responsibility that requires dialogue.

Although the responsibility model does entail a third understanding of nondirectiveness, it shares some of the same concerns as the psychotherapeutic model.[44] Both acknowledge that the counselee's emotional and practical responses cannot be avoided by the genetic counselor. In this way, both are critical of the teaching model's hands-off approach. The HCP undertaking genetic counseling in both models should take an active role in the decision making process. At the same time, both models reject persuasive coercion. Finally, the psychotherapeutic and responsibility models converge on the goal of promoting autonomy in complicated circumstances that call for an act of self-determination.

To begin to understand the distinctiveness of the responsibility model's account of nondirectiveness, one must look to the fundamental vocabularies that express the respective models. The teaching model depends on pedagogical idioms premised on a notion of technical neutrality; the psychotherapeutic model uses a psychological idiom based on a diluted Rogerian notion of psychotherapeutic communication; the responsibility model looks to a normative vocabulary rooted in a pragmatic account of communication. Nondirectiveness on this last view gets its grip by looking at the proprieties of conversational scorekeeping understood within a dialogical model.

As Brandom puts it, scorekeepers keep two sets of books. Scorekeeper A not only acknowledges how a conversation changes A's perspective but also attributes changes to the perspective of scorekeeper B. Discursive moves constrained by inferential proprieties is one cause for these changes.[45] If genetic counseling is understood in light of this model, then the concepts of directiveness and nondirectiveness need to be mapped on to this description. Let us make A the genetic counselor and B the patient.[46] Some conversational moves by A can change the belief set or *doxastic* commitments of B; other moves by A can change the intentions or *practical* commitments of B. And vice versa. The goal of the genetic counselor should be to take directive and nondirective stances during the conversation that are appropriate to the goals of coordinating meanings and facilitating responsible decision making. Directiveness according to the responsibility model involves making discursive moves that *appropriately or inappropriately* direct the scorekeeping activities of the conversation partner(s). A nondirective stance entails discursive restraint that allows someone else in the conversation, in this case the patient, to direct the conversation. Inappropriate directiveness comes in many forms. Kessler's notion of persuasive coercion is one example and can be characterized as intentional misdirection. Scorekeeper A can also be overly directive by giving too much information and not allowing sufficient opportunity for B to direct the conversation. The most common form of inappropriate nondirectiveness is the refusal to direct the scorekeeping despite the patient's request for some help. The rest of the section attempts to demonstrate how these distinctions work.

---

[44] White's and Kessler's views are for instance cited by Weil as advocating nondirectiveness as the promotion of autonomy. See Weil, *Psychosocial Genetic Counseling*, 123.

[45] Other causes that can change scorekeeping perspective are nondiscursive, i.e. nonverbal behaviors.

[46] To help track the perspectives, I will refer to A as male and B as female.

When A undertakes an assertion such as *P* (from Chap. 3), he *directs* B's scorekeeping to a conceptual content about B's risk status. A then proceeds to tell B what *P* means from A's perspective as a genetic counselor. A is directing B to keep score on the inferential significance of *P* from A's perspective as a genetic counselor. At this juncture, A's conversational moves entail a kind of directiveness that is defined much more broadly than Kessler's definition. A is authorized to take this initial directive stance because he is fulfilling the role of genetic counselor, which is assumed to have specialized knowledge about *P*. A proceeds to ask B questions about her understanding of *P*. From A's standpoint, this move is an attempt to shift to a nondirective stance by inviting B to make moves in the conversation. If B requests that A clarify or answer questions, then this prompts A to return to directing B in her understanding but with the important difference that B has explicitly authorized A to continue being directive. In other words, a minimal dialogical process is underway. If B responds to A's invitation by expressing her understanding of *P*, then B is directing A's scorekeeping –as well as her own – to the inferential significance of *P* from her perspective. When *P's* meanings have been coordinated between A and B, A directs B's attention to additional conceptual contents that need to be coordinated. Here again A's role as genetic counselor authorizes him or her to lead the conversation. For the sake of a clear distinction, we will say that the interactions thus far have focused on the doxastic commitments of A and B and have not yet addressed practical commitments regarding decision making. At the appropriate juncture, the conversation shifts to the difference *P* makes on B's practical commitments to take or avoid certain actions. Will *P* prompt B to undergo amniocentesis? When the conversation turns to practical questions, the scorekeeping turns to practical commitments. A directs B's attention to her decision making options and asks B what her preferences are.

This phase of scorekeeping is generally where issues around patient autonomy are focused. Some notions of autonomy confer authority to B because she has individualized and specialized knowledge about her own practical commitments. This reasoning commits A to a nondirective stance because only B knows what she wants. The teaching model subscribes to this view. This reason for nondirectiveness is confounded if B in fact does not have immediate or clear commitments that are relevant to practical reasoning. Genetic counselors experience patients like this on a regular basis, as Kessler's quotation in the introduction indicated. The responsibility model confers decision making authority to the patient for different reasons. These are related to the assumption developed in Chap. 3. It is assumed that the patient wants to and has the capacity to share responsibility for understanding the information and deciding what to do with the information. This way of distributing decision making authority reveals the social dimensions of autonomy.[47] B's practical

---

[47] Hegel is sometimes accused of collapsing autonomy into heteronomy because he claims that autonomy requires that others adopt certain attitudes towards me. Brandom exonerates Hegel with this clarification: autonomy means I get to choose my commitments but you, whoever I am responsible to, holds me to them. See Robert Brandom, *Tales of the Mighty Dead : Historical Essays in the Metaphysics of Intentionality* (Cambridge, Mass.: Harvard University Press, 2002).

commitments are not privileged because she has special access to them; instead they are privileged because of her relationship to the consequences of the decision making. If amniocentesis were undertaken, B's body would have to stuck by a needle; B would have the miscarriage if complications arose from the amniocentesis. If amniocentesis were declined, B would have to manage the anxiety of not knowing; B is responsible for the care of the child to whom she eventually gives birth.

The two primary goals of the responsibility model are to coordinate meaning across diverse perspectives and to facilitate responsible decision making. The first goal should be responsive to the directive structure of the entire genetic counseling enterprise. The offering of genetic information *directs* a patient to make inferences in response to sentences with biomedical vocabulary with which a patient is generally unfamiliar. This kind of directiveness is designated here as *doxastic directiveness* and should not be unconditionally prohibited. Doxastic directiveness entails leading a client towards a certain belief outcome. For example, when a genetic counselor offers a risk assessment it is important to lead some clients to the conclusion that a risk assessment does not mean that they – or the fetus – have the condition. This guidance is the kind of inferential directiveness that clients expect when information is offered to them. To understand a sentence like *P requires* that the client inherit additional premises and conclusions to understand the information from a biomedical perspective. This entitlement to be directive is also a condition for the abuse of inferential power and requires the HCP to be vigilant about biases that might lead to inappropriate doxastic directiveness. Advocates for those with disabilities have criticized genetic counselors for offering high-risk pregnant women only the health concerns related to having a child with Down syndrome.[48] These critics suggest that clients should have access to information about the challenges and *benefits* of raising a child with Down syndrome. Many HCPs omit these benefits without the intent of steering a client towards termination of pregnancy. They are not intentionally misdirecting the patient but rather focusing on what they consider to be the medically relevant features of the condition. According to critics, the HCPs are being insufficiently directive by offering too little information about what it means to raise a child with Down syndrome.

The HCP's perspective is privileged in this allowance for doxastic directiveness. Limited authority must be attributed to the HCP's perspective because it possesses specialized knowledge, but this attribution of authority is not absolute. If meaning is to be coordinated across diverse perspectives, not all doxastic activity should be directed by the HCP. The patient's background commitments should also play a role in grasping what the information means without yet thinking about decision making. The genetic counselor should be nondirective in terms of allowing the patient to

---

[48] Harmon. For more philosophical critiques, see A. Asch, "Distracted by Disability. The "Difference" Of Disability in the Medical Setting," *Camb Q Healthc Ethics* 7, no. 1 (1998), Asch, "Disability Equality and Prenatal Testing: Contradictory or Compatible?.", E. Parens and A. Asch, "Disability Rights Critique of Prenatal Genetic Testing: Reflections and Recommendations," *Ment Retard Dev Disabil Res Rev* 9, no. 1 (2003), Parens and Asch, *Prenatal Testing and Disability Rights*.

make his or her own inferences. Nondirectiveness here is a kind of inferential restraint on the HCP's part. This movement between directiveness and nondirectiveness has the dialogical structure proposed by White's model.

In the second goal, the HCP is charged with facilitating responsible decision making primarily from the perspective of the patient. *Practical* directiveness and nondirectiveness reside in this sphere of conversational scorekeeping. Practical directiveness can also be appropriate and inappropriate. It is inappropriate when the genetic counselor attempts to change the patient score on specific issues. Telling a patient at the outset what their personal preferences should be, or proposing to a counselee what their responsibilities as a parent, patient or human being are is prima facie an unacceptable form of practical directiveness. This kind of directiveness can appear in subtle forms.

Let's take an example from a recent study. In a study of 30,564 pregnancies receiving first-trimester risk assessments,[49] Nicolaides et. al. found a high correlation between increase in risk for Down Syndrome (DS) and increase in consent to undergo invasive testing for diagnostic confirmation. Close to 77 % of the women who had a risk greater than 1/300 chose to have an invasive test and only 4.6 % of pregnant women with a risk less than 1/300 had invasive testing. The researchers concluded that the study "demonstrated that pregnant women are indeed capable of making informed decisions on the basis of the best available evidence."[50] Genetic counselors who associate high quality genetic counseling with a practical nondirective stance might object to telling the women in the study "that although a risk of 1 in 300 or more was *generally considered to be high*, it was up to them to decide in favor or against invasive testing."[51] The italicized content indirectly applies practical pressure by making explicit how most people evaluate the probability. What is implicit is that most people do something about high risks. The researcher could be

---

[49] In October 2003, The *New England Journal of Medicine* reported the results of a large-scale study of first-trimester screening techniques. This report, along with the subsequent endorsement of American College of Obstetrics and Gynecology's in June 2004, ADD January 2007 gives justification for practitioners to incorporate these new techniques into prenatal care. The first-trimester screen combines maternal age, nuchal-translucency thickness, and levels of maternal serum analytes to identify about 78.7 % of fetuses with Down syndrome with a false positive rate of 5 %. The first component involves the measurement of nuchal translucency using ultrasound in the period between 10 and 14 weeks of gestation. If the nuchal fold measures above a certain number of millimeters, then it is considered to be abnormal and possibly associated with Down syndrome as well as other aneuploidies. Access to this screening method may be restricted by the availability of properly trained ultrasound technicians. It is not considered accurate without the second component of the first trimester screen involving the measurement of two biochemical markers: free beta subunit of human chorionic gonadotropin (ß-hCG), and pregnancy-associated plasma protein-A (PAPP-A). At specified levels, these markers are associated with an increased risk of trisomy 21 and trisomy 18. Unlike second trimester noninvasive techniques, first trimester methods do not calculate risk for spina bifida, the most common neural tube defect. Screening for neural tube defects must be done in the second trimester.

[50] K. H. Nicolaides and others, "Evidence-Based Obstetric Ethics and Informed Decision-Making by Pregnant Women About Invasive Diagnosis after First-Trimester Assessment of Risk for Trisomy 21," *Am J Obstet Gynecol* 193, no. 2 (2005): 324.

[51] Ibid., 323.

accused of smuggling in a commitment to be used in practical reasoning that leads to undergoing an invasive test. Practical nondirectiveness is the stance by the genetic counselor that allows the patient to ultimately decide what evaluations, preferences, or obligations are relevant in this circumstance.

There are appropriate forms of practical directiveness as well. When a patient has stated emotional reactions, personal preferences or moral obligations in response to their options, a genetic counselor can help direct *provisional* inferences using the commitments from the patient's perspective. For example, if a pregnant woman stated that she believed abortion was wrong in most circumstances, then a genetic counselor could help her think about the how this belief might function as a reason not to undergo amniocentesis in certain circumstances. In other words, the HCP can try to assist the patient in making the right decision from the patient's perspective. One should notice that practical nondirectiveness privileges the patient's perspective but, similar to the HCP perspective in a doxastic dimension, the patient's perspective does not have absolute authority.

White discusses this possibility for directiveness in terms of the genetic counselor's role as gatekeeper.[52] A genetic counselor can challenge a preference or obligation of a patient that is incompatible with the HCP's role or widely accepted social norms. Such challenges will be rare but are on occasion justified. Several cases have been raised to illustrate the difficulties of these situations in a pluralistic society. If a pregnant woman in the U.S. indicates she wants a child but not a girl for cultural reasons and decides to terminate upon finding out the sex of the fetus, should the HCP doing genetic counseling challenge this culturally backed preference? Is this incompatible with a widely held U.S. norm against sex discrimination? If a deaf couple decides to terminate a fetus after finding that he or she will hear normally, should a genetic counselor challenge this preference for a child to fit the norms of the Deaf culture? These are difficult situations that directly confront a counselor's commitment to practical nondirectiveness.

On the responsibility model's view, these are circumstances that involve real conflicts that must be negotiated across diverse perspectives. On the one hand, White is right to limit the genetic counselor's authority to challenge the patient in a pluralistic society committed to maximizing individual freedoms. For example, if the HCP's own religious commitments were incompatible with patient preference, then this disagreement would not entitle the genetic counselor to challenge the patient's preferences. If these kinds of conflicts arise consistently for the HCP, then an internal negotiation must occur between HCP's stance toward his own role obligations and religious obligations. On the other hand, the genetic counselor does have some bases to disagree with the patient. The political and medical norms that are established to guide the use of genetic information are legitimate constraints on patient preferences. An example of a political norm is the regulatory structure around termination of pregnancies; an example of a medical norm is the professional commitment to avoid doing harm to patients. Because the norms concerning

---

[52] White, "Making Responsible Decisions. An Interpretive Ethic for Genetic Decisionmaking," 19.

the medically appropriate uses of genetic information are not fully established, the cases above demonstrate a kind of case law approach to the formation of norms.[53]

The responsibility model of nondirectiveness creates distinctions to understand the variety of directive and nondirective action that take place in a genetic counseling session. All of these are understood in a normative framework that acknowledge both doxastic and practical inferences are part of conversational scorekeeping This does not mean that objectivity has no status; instead it shows that even in situations where standardizable knowledge is offered, there are right and wrong ways of keeping score both as a HCP and a patient. And given that normative appraisal is in play, one must acknowledge that in certain moments the HCP's perspective has privileged but not absolute authority, and in other moments the counselees perspective is privileged even if open to challenge. This stance avoids on the doxastic side the view that genetic information means whatever the patient says it means; at the same time its meaning is not wholly determined by the biomedical perspective. In the realm of practical commitments, the responsibility model acknowledges that a HCP must defer to the perspective of the patient to generate preference or obligations unless the patient's perspective endorses commitments that are incompatible with firmly established professional or societal norms.

Evaluation of Models: Debbie's Case

In the introduction, I presented Debbie's case highlighting several issues that deserve attention. One concern was the question of whether the genetic counselor should have offered to leave the room during an intense moment of deliberation. This question along with other aspects of the case relevant to nondirectiveness will be addressed here as a way to evaluate the different understandings presented above.

The development of Debbie's case in many ways was a textbook example of the teaching model. The risk assessment is presented in a clear methodical fashion, and patient misconceptions about amniocentesis are corrected. When Debbie expresses concern over putting the baby at risk and concern about leaving her present children with care giving responsibilities, the counselor tries to help her think about the situation by listing each of the practical outcomes. The genetic counselor observes that Debbie is ambivalent and may need privacy to discuss emotional and religious matters.

From the standpoint of the teaching model this counseling session upholds nondirectiveness in three ways. First, the genetic counselor does not try to direct Debbie's decision making when it is clear that she is ambivalent. This nonintervention is justified by the fact that the genetic counselor cannot understand Debbie's perspective especially her conflicting preferences and religious concerns. This gap in understanding is consistent with the teaching model assumptions about

---

[53] For an example of a proposal to codify these norms, see J. R. Botkin, "Federal Privacy and Confidentiality," *The Hastings Center Report* 25, no. 5 (1995).

communication. In Chap. 2, I attempted to show that the teaching model adopts the technical vision of communication and understandings of technical neutrality. In this view objective information can be transmitted cleanly whereas subjective information has less probability of being transmitted because of its indeterminacy. In semantic terms, the genetic counselor has offered Debbie the objective meanings and options and leaves the subjective meanings to the proper authority, Debbie and her husband.

The psychotherapeutic model would have several criticisms of how the genetic counselor adhered to the value of nondirectiveness, if understood as promoting autonomy. First, Debbie's counselor failed to acknowledge the basic need of the client to be validated as a decision maker. No indication is given that the genetic counselor is assessing or articulating Debbie's strengths as a decision maker, only the implicit trust conveyed in leaving the Debbie alone to decide. The genetic counselor does not undertake follow-up questions about the concerns regarding responsibilities, risk and religion. This omission is a failure to attempt empathic identification that allows the HCP to see the situation from the counselee's standpoint. Ignoring these issues also potentially conveys the message that these issues are too difficult or scary to address. Finally, the offer to leave the room to allow time for Debbie and her husband to discuss might be interpreted by the client as abandonment during a difficult moment. From the psychotherapeutic standpoint, this departure is a breach of a perspectival partnership where the counselor and the client together try to interpret and act on the genetic information from the client's perspective.

How should the HCP have proceeded in Debbie's case in terms of the psychotherapeutic notion of nondirectiveness? If nondirectiveness involves interventions that promotes the patient's autonomy, then there are numerous actions that could have been taken and they fall into two basic strategies: (1) understand the patient (2) support and build patient confidence in making decisions. The counselor in this case should have spent much more time trying to understand Debbie's concerns about leaving her children with care giving responsibilities. The HCP should have tried to discern how Debbie understood the probability of miscarriage and openness to terminating the pregnancy. She should have also asked Debbie more about her religious commitments, a topic that will be taken up in full in the next chapter. All of these strategies provide the counselor access to Debbie's 'interior' and allow assessment of Debbie's strengths and weaknesses. With this information, the counselor could have made a variety of comments to build Debbie's self-esteem. For example, after Debbie indicates the concern about leaving her children with responsibilities, the genetic counselor could say: "You are a thoughtful mother who clearly cares about her children.' Or after discussing the religious content, the HCP could say, "You are clearly a spiritual person who wants to do right in the eyes of God."

The responsibility model also has several criticisms of how Debbie's case was handled and by association criticisms of the teaching model's notion of nondirectiveness. Remember that the responsibility account acknowledges the need for movement between directiveness and nondirectiveness in both the doxastic and practical aspects of the conversation. First, there seems to be minimal undertaking of doxastic nondirectiveness on the counselor's part. If Debbie's response to the risk assessment was a nod, then the genetic counselor should have tried to elicit more thoughts and reaction from Debbie about her understanding of what was said. Although eliciting Debbie's thoughts is a directive act, the request intends to promote doxastic nondirectiveness by creating space for Debbie to share her feelings and inferences. If Debbie has no comment, then the difficulty of sorting through the numbers needs to be acknowledged and another explanation possibly needs to be given. Also, if an emotional reaction is observable, then this needs to be acknowledged as part of the meaning of the information: "What I am telling you may be very difficult to hear." This could allow Debbie space to connect her feeling to her understanding of what she heard. Doxastic nondirectiveness in this case allows time for the meaning of the genetic information to be connected to the emotional states that it causes. Debbie's expression of God's will should have also been allowed more time to develop as an inferential pathway to understanding the situation.

Second, the practical nondirectiveness enacted by the genetic counselor should have been supplemented by practical directiveness. Debbie offers practical concerns that the counselor should have helped her develop. For example, the genetic counselor could have asked what it would mean to Debbie to lose this baby in a miscarriage and does the risk sound high or low in relation to such a loss. Like the psychotherapeutic model, the genetic counselor guided by the responsibility model would have to receive clear signals from the client to extend an offer to leave. Whether such preferences were expressed by Debbie's body language are not indicated in the case.

In evaluating the three models via Debbie's case, it is easier to identify how the responsibility model differs from the teaching model. The teaching model has two modes, 1) educate the client, what the responsibility model calls doxastic directiveness 2) refrain from giving advice or do not undertake practical directiveness. In the teaching model, the meaning of the genetic information is overdetermined by the HCP's perspective and underdetermined by the patient's perspective. In terms of the decision making process, the patient's perspective should be privileged, but that does not mean that the patient has complete access to her own situation. The justification for this criticism is that no perspective has that much semantic control or grasp of meanings either in terms of understanding or decision making. The meaning of the genetic information is not totally up to the genetic counselor; at the same time, the patient does not have a full grasp of her own background commitments in relation to the genetic information. The teaching model's account of nondirectiveness is part of a larger problematic picture of communication as transmission. Once the genetic information is transmitted objectively then a rational person needs freedom to make logical inferences to make a decision. The problem with this picture is that communication and decision making rarely work in this tidy two-step pattern. The responsibility picture of nondirectiveness attempts to capture the more difficult conversational "scorekeeping kinematics."[54]

---

[54] Brandom, *Making It Explicit : Reasoning, Representing, and Discursive Commitment*, 142.

Evaluation of Models: Debbie's Case

A more difficult task is to compare the responsibility model to the psychotherapeutic model's notion of nondirectiveness. Kessler narrows directiveness to persuasive coercion and expands the conceptual realm between and within nondirectiveness. In Debbie's case this means that there is no directiveness that occurs and on Kessler's conception of nondirectiveness, the genetic counselor fails to promote Debbie's autonomy in any way above providing information. As Kessler suggests this leaves a lot of interaction that cannot be sorted into the directive-nondirective slots, but I think this misses too much.

The responsibility model seeks to articulate sites of authority and the patterns of inferential activity as they change throughout the conversation. As a result, doxastic and practical directiveness are modes that are appropriate in some circumstances and not in others, and this depends on the conversational score and which locus of authority is privileged at that moment. All the models agree that persuasive coercion is never appropriate but to define directiveness only in terms of this sacrifices its role in discerning the normative balance that is needed between the directive and nondirective moments in the genetic counseling conversation. For example, the perspective that is privileged depends on the score at that moment in the conversation; and ascriptions of directive/nondirective action depend on who is actually doing the inferential or emotional work at what moment. Kessler's definition of directiveness misses this dimension of the exchange entirely. As a result, his conception of nondirectiveness casts too large a net making it difficult to make important distinctions about how *directive* and *nondirective* actions promote autonomy. He does acknowledge these distinctions, but they are difficult to fit within his model. For instance, his model calls for active counseling skills that are directive actions. But he is then put in the semantic bind of needing to call directive actions nondirective. The appropriate moments for doxastic and practical directiveness in the responsibility model are for Kessler nondirective actions.

At a more fundamental level, the responsibility model trades in talk of psychology and validation for normativity and responsibility. This difference is indicated in the different conceptions of nondirectiveness. In the psychotherapeutic model, nondirectiveness is rooted in the assumption that all clients need validation and need to be understood by the genetic counselor; whereas on the responsibility view all clients are sites of authority with which meaning needs to be coordinated and decisions needs to be facilitated. Whether a client's confidence in decision making needs boosting is a difficult assessment to make in a short session, but a counselor can safely assume that patients might need some help taking responsibility for grasping the full implications of the genetic information and for the decision that needs to be made.

The objections to the responsibility notion of nondirectiveness are several. I will address two important ones. From the teaching standpoint, the concern should be that HCPs will abuse their entitlement to be practically directive. If genetic counselors help counselees make practical inferences from stated preference or obligations, the genetic counselor can subtly manipulate the conclusions drawn. This manipulation is a possibility but the model has a guideline on this matter. If the counselor draws a practical inference on behalf of the client, e.g., 'given X, then maybe you are not comfortable undergoing amniocentesis,' the resulting conclusion must always be cast as a provisional suggestion for a practical pathway that can be challenged by the client. Also, the genetic counselor should be explicit that she is helping the patient think about the consequences of her own stated values.

From the psychotherapeutic standpoint, the responsibility model's understanding of the directiveness/nondirectiveness distinction will not fully address the psychological needs of the patient. The responsibility model's concern towards the issues of authority and responsibility promotes a sensitivity to a counselee's status as someone who needs to explore meaning and make decisions, who needs to be directed in some parts of the conversation and who needs to direct at other junctures. A counselee whose emotions consistently halt the conversation or whose psychological state appears to prevent an adequate grasp of the situation may not in those moments or session be able to take responsibility for meaning or decisions. Making that conclusion and referring that person for further help easily falls within competencies of the genetic counselor guided by the responsibility model. As for dealing with strong emotional reactions by patients to genetic information, the responsibility model does not presume this is a psychological problem but rather a common result of the information that needs to be incorporated into its meaning and the dialogical process. Words affect meaning and bodies.

Summary

In this chapter, I have addressed one of the central values of genetic counseling. Acknowledging its indeterminate status and history, I then elaborated it from the three models under consideration and tried to demonstrate why the responsibility model is the most sufficient. The responsibility model proposes that directiveness and nondirectiveness are inevitable features of genetic counseling and that the ability to discern when these are appropriate is a crucial skill. The distinction between doxastic and practical directiveness/nondirectiveness acknowledges the boundary between meaning making and decision making but makes it permeable in the sense that the genetic counselor can be directive and nondirective in both areas under certain conditions.

I raised a question at the beginning whose answer has remained implicit: whether nondirectiveness should remain an integral part of the profession. In light of the responsibility model's reframing of the nondirective/directive question, clearly *both* directiveness and nondirectiveness are needed to identify the changing loci of authority and activity in a conversation and to track who is doing what inferential work in the genetic counseling session. In the next chapter, the need for these concepts will be further demonstrated in the concern around the issues of talking about spirituality and religion.

Chapter 5
Genetic Counseling and Spiritual Assessment

Or, if, as we suspect based on their attitudes toward abortion, religious faith were important to the couple, (a clinician could say) this:

God sometimes gives people special tasks in life that we need to do without rewards or thanks and yours is to care for Alexis until He's ready to take her. And, the two of you have done a magnificent job in caring and loving her. You may not experience her love now, but a day will come when you will. God will take care of that. (Seymour Kessler and Robert Resta in "Commentary on Robin's *A Smile*, and the Need for Counseling Skills in the Clinic"[1])

In the previous chapter, the value of nondirectiveness was shown to be a contested and defining feature of genetic counseling. It was interpreted from the standpoint of the teaching, psychotherapeutic and responsibility models and then respectively applied to Debbie's case. The responsibility model proposes that genetic counselors need to acknowledge the presence of directiveness and nondirectiveness in both meaning-making and decision-making processes. A key communication skill in genetic counseling –and all health care communication for that matter – is recognizing when directiveness or nondirectiveness is appropriate based on the current conversational score. One area where this skill is difficult to employ involves issues of spirituality and religion.

If patients bring up spiritual or religious concerns in response to receiving genetic information, should the genetic counselor help the counselee understand the genetic information in religious terms? This question can be taken one step further. If some studies show that many patients want their spiritual concerns addressed in health care situations or if they show that acknowledging this aspect of patients' lives improves health outcomes, then should genetic counselors formally assess a patient's spiritual needs and concerns? Are spirituality and religious concerns different from other psychosocial factors that genetic counselors are trained to handle? As Kessler exhibits in the above quotation, some genetic counselors are comfortable

---

[1] R. G. Resta and S. Kessler, "Commentary on Robin's a Smile, and the Need for Counseling Skills in the Clinic," *Am J Med Genet A* 126, no. 4 (2004): 437.

articulating a patient's or parent's situation in religious terms. Studies have found that this is a minority position.[2]

This chapter addresses two questions. The first asks whether spiritual assessment should be a routine part of genetic counseling sessions. Some researchers in genetic counseling as well as in medicine are investigating the feasibility of adopting standard spiritual assessment tools. After reviewing their findings, I will offer a challenge to the position that spiritual assessment should be routine part of genetic counseling session. The second question asks how the religious concerns in Debbie's case should have been handled. After interpreting this case through the lens of each of the three genetic counseling models, I argue that responsibility model provides the most adequate understanding of and guidance for this area of interaction.

Spiritual Assessment in Genetic Counseling

Health care professionals are paying more attention to the spiritual lives of patients.[3] This area of research can be divided into three main areas: (1) spirituality and patient/HCP interest (2) spirituality and health outcomes (3) spirituality and patient care. The findings from this research are being used to justify inquiries into the feasibility of instituting formal spiritual assessment tools. These findings as they appear in the genetic counseling literature will be assessed in this section but prior to this work, an assessment of a representative definition of spirituality and a subsequent proposal for a more comprehensive definition are undertaken.

Defining Spirituality

In this section, a representative definition of spirituality used by researchers in medicine and genetic counseling is introduced and analyzed. The shortcomings of this definition warrant the proposal of more comprehensive definition. I offer a

---

[2] For empirical studies that show hesitance of genetic counselors to engage in this activity, see L. M. Reis and others, "Spiritual Assessment in Genetic Counseling," *J Genet Couns* 16, no. 1 (2007). and P. J. Boyle, "Genetics and Pastoral Counseling: A Special Report," *Second Opin (Chic)*, no. 11 (2004).

[3] One rough indicator of the spate of interest in this area is that 2283 of 2512 medical publications on "spirituality" and "medicine" have occurred in the last 15 years. These numbers are based on a key word search of "spirituality"and"medicine" in Pubmed, a large research database provided by the National Library of Medicine and the National Institutes of Health. An interesting historical and sociological question not pursued here is why there has been such a surge interest. The ethical question of the benefit and harms of the relationship between religion and medicine has been addressed. See Richard P. Sloan, *Blind Faith : The Unholy Alliance of Religion and Medicine*, 1st ed. (New York: St. Martin's Press, 2006).

definition of spirituality that captures a wider range of attitudes that patients might exhibit in genetic counseling.

A dominant view within the medical literature identifies spirituality as the genus of which religion is a species.[4] In most health care contexts, the concern is how to define spirituality as the umbrella concept that includes religion as well other related phenomena. Of the many accounts offered,[5] Anandarajah and Hight propose a representative definition:

> Spirituality is a complex and multidimensional part of human experience. It has cognitive, experiential, and behavioral aspects. The cognitive or philosophic aspects include the search for meaning, purpose and truth in life and the beliefs and values by which an individual lives. The experiential and emotional aspects involve feelings of hope, love, connection, inner peace, comfort and support. These are reflected in the quality of an individual's inner resources, the ability to give and receive spiritual love, and the types of relationships and connections that exist with self, the community, the environment and nature, and the transcendent (e.g. power greater than self, a value system, God, cosmic consciousness). The behavior aspects of spirituality involve the way a person externally manifests individual spiritual beliefs and inner spiritual state.[6]

This definition has relevance for the present study because it is offered in conjunction with the HOPE spiritual assessment tool, an instrument that guides clinicians in asking patients a series of questions about their spiritual lives and that has been considered in genetic counseling research. One key function of this definition is that it seeks to facilitate a HCP's ability to recognize and distinguish the spiritual concerns of patients. It identifies both religious and nonreligious forms of spirituality allowing for a diverse set of phenomena that could include weekly communal worship, private meditation, hiking or political activism, any of which can influence how individuals respond to a wide range of medical situations. The definition acknowledges that spirituality is a phenomenon that influences the way patients think, feel, behave. The cognitive dimension of spirituality is identifiable in what is said and thought about the broader or transcendent circumstances and consequences of individual and collective existence. The experiential or emotive dimension of

---

[4] A large literature in the humanities addresses the distinction between spirituality and religion. Medical researchers do not pretend to have sorted through the thicket of these conceptual nuances. They have adopted a stipulative definition that serves their purposes. Within the humanistic tradition, I think the biomedical community's preference is most compatible with a Hegelian understanding of the distinction that he worked out in Hegel, Miller, and Findlay., and in Georg Wilhelm Friedrich Hegel and Peter Crafts Hodgson, *Lectures on the Philosophy of Religion : The Lectures of 1827*, one-volume edition. ed. (Berkeley: University of California Press, 1988).

[5] To review other definitions of spirituality in the medical literature, see J. Dyson, M. Cobb, and D. Forman, "The Meaning of Spirituality: A Literature Review," *J Adv Nurs* 26, no. 6 (1997); A. Moreira-Almeida and H. G. Koenig, "Retaining the Meaning of the Words Religiousness and Spirituality: A Commentary on the Whoqol Srpb Group's "A Cross-Cultural Study of Spirituality, Religion, and Personal Beliefs as Components of Quality of Life" (62: 6, 2005, 1486–1497)," *Soc Sci Med* 63, no. 4 (2006).; W. McSherry and K. Cash, "The Language of Spirituality: An Emerging Taxonomy," *Int J Nurs Stud* 41, no. 2 (2004).

[6] G. Anandarajah and E. Hight, "Spirituality and Medical Practice: Using the Hope Questions as a Practical Tool for Spiritual Assessment," *Am Fam Physician* 63, no. 1 (2001): 83.

spirituality is recognizable in generally *positive* feelings or psychological states, i.e. hope, love, inner peace, about relationships and circumstances. The expression of these beliefs and inner states in observable behaviors articulates the last dimension that refers to the influence of spirituality on action.

This definition has several strengths. First, it avoids reducing spirituality to an essentially emotive phenomena marked by pseudo-claims or superstitious acts. Whether this avoidance leads to an over-intellectualization will be discussed below. Second, it avoids reducing spirituality to codified or institutionalized beliefs. The cultural power of institutionalized religious traditions – at least in parts of the U.S. – sometimes leaves other forms of spirituality at the margin of our awareness. Most clinician researchers interested in spiritual assessment have consistently included nontraditional religious expression in their stipulative definitions. Finally, an HCP guided by this definition will grasp the complexity of spirituality and expect it to be expressed in a variety of ways. Patients may express their spirituality as a belief, "I believe that things happen for a reason;" or as a report of an attitude, "I'm hopeful;" or as an action such as walking to the hospital chapel.

The definition has shortcomings as well. Its avoidance of emotivist or essentialist descriptions results in an implicitly intellectualized phenomenon. The relationship between cognition, emotion, and behavior imply a trickle down spirituality where beliefs influence emotional states and emotional states produce certain kinds of behaviors. The relationship between these dimensions is more complex than the definition implies. A second weakness is its representation of the relationship between experience, emotion and feeling. It suggests that experiential and emotional aspects of spirituality produce a set of feelings. The normative categories of dispositions or attitudes provide a more accurate description of these phenomena. The third and most important weakness of the definition is its attenuation of the phenomena associated with spirituality. The authors acknowledge that the absence of one or several features can result in a spiritual crisis. This reference to spiritual crisis indirectly reveals the weakness. It denotes spirituality as the *presence* of a set of features without acknowledging that spirituality is also the realm where these same attitudes, emotions, and behaviors can be absent or hindered. The beliefs, attitudes and emotions that are mentioned can only be understood in connection to other attitudes such as despair, uncertainty, and fear. For example, the Spirit leading Jesus into the wilderness is not often described as a spiritual crisis but a time of spiritual growth through trial. Or despair after receiving a devastating genetic diagnosis would not necessarily be the absence of spirituality but possibly ingredient to spiritual practice. Spirituality in the above definition substitutes one kind of stance or achievement within spirituality, an exemplary one, for the cultural mode of spirituality defined by circumstances where attitudes of hope, despair, love, sadness, joy, uncertainty and faith are lived, acknowledged or called upon to in response to difficult situations.

These definitional deficiencies prompt the need for a more comprehensive definition of spirituality. Daniel Sheridan proposes a way of working out the distinctions between culture, spirituality and religion that proves helpful both for students of religion and for the biomedical community. He defines spirituality as "a mode of culture in which the human being transforms the problematic of the human predicament immanently within the plenum and spectrum of human resources in time and space."[7] One innovation in this definition is that culture plays the functional role often ascribed to religion, and spirituality becomes a substantive domain within cultures.[8] This makes culture the genus of which spirituality is a species:

> Culture is an open, complex, systemic whole of human behavior acquired and transmitted by symbols, constituting the distinctive achievement of human groups. The essential core of culture consists of traditional ideas and values. A culture is both a product of action and conditions further actions. Culture has a function, among others, to transform the human predicament, that is, the inherent dilemmas of being finitely human in time and space, possibly in the assisting presence of a Transcendent Other.[9]

Spirituality as a mode of culture or subculture can be further distinguished between what he calls plenum spirituality and axial spirituality. Plenum spirituality refers to the context of transformation as occurring within the whole, a monistic totality where everything is interconnected. In this spirituality, gods, humans, nature are inextricably bound together in a cosmic whole. Axial spirituality emphasizes human resources as distinct from the whole to transform its own problematic. The varieties of humanism exemplify but do not exhaust this category of spirituality.

After distinguishing plenum and axial spiritualities, religion is defined as a similar mode of culture with the addition of "the presence of an assisting Transcendent Other, which is perceived to be without space and time."[10] For Sheridan, spirituality is not a religion that lacks a transcendent dimension; instead, religion is a spirituality that adds a theistic concern. "In this framework the Transcendent Other is seen to be 'additional,' almost in a sense not necessary since certain cultures, for example, Theravada Buddhism, do not postulate it."[11] He illustrates how all three concepts can be present within single traditions or cultures:

---

[7] Daniel Sheridan, "Discerning Difference: A Taxonomy of Culture, Spirituality, and Religion," *The Journal of Religion* 66, no. 1 (1986): 43.

[8] Sheridan's proposal is an attempt to resolve the problem of defining religion and culture as different phenomena. When religion is defined broadly, i.e. Geertz's religion as cultural system, it becomes difficult to distinguish it from the concept of culture. See Clifford Geertz, *The Interpretation of Cultures; Selected Essays* (New York,: Basic Books, 1973), 87–125.

[9] Sheridan: 40. Sheridan specifies several features of culture: "Underlying this conception of culture are four implied factors: (1) a cosmology, (2) a view of the problematic of the human predicament, (3) a goal of transformation of the predicament, and (4) specific means of transformation of that predicament. The four factors describe a worldview."

[10] Ibid., 44. His reference to plenum suggests spiritualities that are transformed without distinctions such as culture/nature. His "axial spirituality" is an allusion to Karl Jaspers description of the axial period (800–200 B.C.) of pivotal thinkers that transformed humanity's self-understanding.

[11] Ibid., 45. Sheridan's taxonomy has some unsatisfactory consequences as do most stances at this level of categorization. The most important has to do with his demarcation of religion. By narrowing it to those modes of culture that look to a transcendent Other for assistance, some phenomena, e.g. most forms of Buddhism, that are traditionally referred to or self-ascribed as religions no longer fall under this mode of culture. Such phenomena would be more likely fall under plenum or axial spiritualities. This change in classification would not put Buddhism or Confucianism outside the purview of scholars of religion. Instead, Sheridan claims that it "obviates the dilemma of trying

Catholicism includes a "little tradition" akin to plenum spirituality, a "great tradition" akin to religion, and a "liberal" tradition akin to axial spirituality. Hinduism includes polytheisms akin to plenum spirituality, yoga akin to axial spirituality, and Vaishnavism akin to religion. Twentieth-century Europe included Nazism akin to plenum spirituality, Barthian neo-orthodoxy akin to religion, and Leninist Marxism akin to axial spirituality.[12]

His taxonomy helps justify the preference in the medical literature for using spirituality as a more general cultural phenomenon than religion. The challenge is to translate these sociological insights into a more stipulative definition for HCPs who are trying to identify the spiritual concerns of individual patients.

When HCPs attempt to the identify spirituality, I propose they look for spirituality in the intentional stances of patients that transform the limiting conditions of their knowing (e.g. cognitive uncertainty), doing (e.g. practical/moral uncertainty) and being (constraints of embodiment). Sheridan's work shows that intentional stances are not de novo creations of individuals but rather these stances are moves within trajectories of meaning bound by the historical and social possibilities/emphases of a culture but not completely determined by them. An individual forges meaning out of those possibilities. In a pluralistic society, HCPs should not be surprised to find patients that exhibit one or more these modes of culture. This helps explain how an individual patient can avow that all available medical interventions should be used (axial spirituality) and that God's will is in every outcome (religion).

Intentional stances are not reducible to beliefs and do not require self-consciousness. An intentional stance can be attributed because of something a person says, 'This disease is God's will,' or it can be attributed to actions a person undertakes, e.g. bowing one's head in prayer or taking a hike in the woods. In terms of emotions, intentional stances – either belief or action – are related to emotions by either trying to affect future emotional responses, e.g. belief in afterlife mitigates despair; or by responding to present or past emotional responses, e.g. praying with the rosary is response to a present anxiety over uncertainty. Intentional stances cannot be reduced to the physiological states we call emotions but they can be 'lodged in emotions' through our efforts to discipline – prospectively or retroactively – this

---

to show that Confucianism or Nazism is a religion, a quasi-religion, a philosophy of life, or an ideology. On this level of interpretation, the scope of the 'study of religion' is universal both synchronically and diachronically."(p. 44–45). In other words, 'religious studies' would be a form of cultural studies that approaches cultures as consisting of plenum/axial spiritualities and/or religions. The benefits and costs of this approach are related to the large net it casts for students of religion. One benefit is that it permits or invites comparison of cultural modes that seek to transform the human predicament. Relevant to this project, medicine and specific religions could be compared as distinct subcultures that seek to transform human health predicaments. One cost of Sheridan's taxonomy is the new set of boundary issues that concern whether a mode of culture actually seeks to 'transform the human problematic' or merely, for example, seeks to profit or distract from the human predicament. The complex circumstances and consequences of Sheridan's proposal cannot be rehearsed here in their entirety. As a stipulative prospect for understanding religion and religious studies, his taxonomy has promise at certain theoretical levels.

[12] Ibid.

responsive feature of our bodies.[13] Intentional stances can be attributed as dispositions if their reliability is observed over time.

Limiting conditions are the features of the human predicament that restrict our capacities to respond to the world.[14] Death, disability, difference, suffering, evil and uncertainty are examples of limiting conditions. The importance of a limiting condition can be evaluated in terms of the comprehensive effects of the constraint on a person/group/environment and the effort required to accept or overcome the limitation. Death for example is a limiting condition that has generated a plurality of complex intentional stances from religious narratives and ritual practices to DNRs and life support. From the standpoint of the person facing death, it has the totalizing effect of ending bodily existence and it is ultimately impossible to overcome. As a limiting condition death ranks high as does suffering in its many forms. Uncertainty about what is the case, what will be the case and what should be case marks off another set of important limiting conditions whose relevance will be discussed in reference to Debbie's case. What it means to transform these conditions depends on the possibilities of spirituality within a culture. Some spiritualities may emphasize the acceptance rather than the avoidance of death; other spiritualities may try to overcome it at all costs.

One advantage of appropriating Sheridan's work is that spirituality is defined broadly enough to locate not only religion but also medicine. Both are subcultures. If religion is a spirituality that is concerned with the assistance a Transcendent Other, then western biomedicine is a spirituality that is concerned with transformation of morbidity and mortality in the presence of scientific evidence and technology. Medicine fits squarely but not exclusively within axial spirituality. Its concern with the relation between religious practices and health outcomes, e.g. praying for or with patients, and alternative modes of healing e.g. naturopathic medicine,[15] suggest the possibility of finding religion and plenum spirituality within the larger biomedical tradition. The benefit of locating both phenomena within spirituality is that it promotes a greater awareness of the HCP's location as he or she ventures to undertake a spiritual assessment.

The constraints of this project do not allow a full defense of this proposal for defining spirituality. A main objection to it might be that it defines spirituality more broadly than common usage allows. Any subculture that "transforms the problematic of the human predicament" is a spirituality. All definitions of spirituality must attempt to navigate between grasping too much and too little. The one appropriated

---

[13] In informal conversation and unpublished documents, Larry Churchill has used this image of beliefs lodged in emotions.

[14] Thomas F. O'Dea, *The Sociology of Religion*, Foundations of Modern Sociology Series (Englewood Cliffs, N.J.,: Prentice-Hall, 1966). The notion of a limiting condition or its function equivalent has a long history in the sociology of religion. O'Dea's work provides an example of a functionalist account of religion: "Thus functional theory sees the role of religion as assisting men to adjust to three brute facts of contingency, powerlessness and scarcity (and consequently, frustration and deprivation)."

[15] See *Vanderbilt Center for Integrative Health*, (Vanderbilt University Medical Center, 2007, accessed December 17 2007); available from http://www.vanderbilthealth.com/integrativehealth/.

here takes responsibility for its expansiveness by identifying further distinctions within spirituality that allow different features of various subcultures to be recognized. This recognition in turn allows HCPs to see patients as having complex intentional stances toward medical care that reflect several modes of cultural influence.

Initial Motivations for Spiritual Assessment

Despite challenges in defining spirituality, medical researchers, relying on the assumption that people generally know what spirituality and religion is, have found that many patients in the U.S. have spiritual/religious needs in medical circumstances and that some want health care professionals to address these concerns.[16] Researchers interested in instituting spiritual assessment in genetic counseling refer to these kinds of studies as motivations for their efforts. In a survey about spiritual assessment in genetic counseling Reis et al. refer to three studies that indicate patient interests. One finds that 77 % of seriously ill patients think physicians should consider their spiritual concerns[17]; a second study finds that 53 % of seriously ill patients feel physicians should *discuss* spiritual needs[18]; a third study of adult outpatients finds that 66 % of respondents believed that inquiries by physicians into their spiritual lives would strengthen the trust in the relationship.[19] These results are far from conclusive, but they are significant because they serve as reasons to explore the possibility of doing spiritual assessments in genetic counseling.

The second area of research that motivates interest in spiritual assessment involves investigations of the correlations between the spiritual activities of patients and health outcomes. This line of inquiry asks questions not only about the effects that spiritual activities, e.g. prayer, have on the natural history of disease but also the effects that religious beliefs have on medical decision making, which ultimately has bearing on health outcomes. Citing nine studies, Reis et al. interpret the cumulative effect of their results:

> Research in the past 20 years increased our understanding of the connections between spirituality, religion and health. Spirituality has positive effects on mental, physical, and

---

[16] Some studies cited by Anandarajah are J. W. Ehman and others, "Do Patients Want Physicians to Inquire About Their Spiritual or Religious Beliefs If They Become Gravely Ill?," *Arch Intern Med* 159, no. 15 (1999), D. E. King and B. Bushwick, "Beliefs and Attitudes of Hospital Inpatients About Faith Healing and Prayer," *J Fam Pract* 39, no. 4 (1994).; T. A. Maugans and W. C. Wadland, "Religion and Family Medicine: A Survey of Physicians and Patients," *J Fam Pract* 32, no. 2 (1991), O. Oyama and H. G. Koenig, "Religious Beliefs and Practices in Family Medicine," *Arch Fam Med* 7, no. 5 (1998).

[17] L. C. Kaldjian, J. F. Jekel, and G. Friedland, "End-of-Life Decisions in Hiv-Positive Patients: The Role of Spiritual Beliefs," *Aids* 12, no. 1 (1998).

[18] King and Bushwick.

[19] Ehman and others.

emotional health including coping ability, self esteem, and social support systems. In addition religious and spiritual beliefs can profoundly influence medical decision making.[20]

The consequence of these findings appears self-evident: HCPs need to address this dimension of patients to improve health outcomes. According to Harold Koenig, a leading scholar in the area of health care and spirituality, this dimension of patients' lives is as important as their psychological state or social circumstances. Arguing for spiritual assessment, he states three goals of spiritual assessments: (1) learn the religious beliefs of patients especially as they pertain to medical care (2) understand how these beliefs influence coping with illness (3) establish the spiritual needs of the patient.[21] The possible benefits of achieving these goals are many. Spiritual assessment "communicates respect for patients spirituality," "obtains information support system" and "may enhance the patient's coping, influence patient compliance and identify individuals who may benefit from a referral for pastoral counseling."[22] These benefits and specified goals along with the studies that support them have led genetic counselors to explore the feasibility of spiritual assessment. Before turning to a detailed explication of two studies in genetic counseling, the arguments against taking a spiritual assessment or spiritual history need to be rehearsed.

Most of the conclusions above have been challenged in Richard Sloan's *Blind Faith: The Unholy Alliance of Religion and Medicine.* Calls for spiritual assessment in response to patient interest are based on questionable research according to Sloan. Through analysis of several important studies, he demonstrates that these studies are methodologically flawed and their conclusions spurious or at best suspect. They tend to overstate patient interest and recommend spiritual assessment for most medical circumstances despite that patient reports are often from a narrower range of medical circumstance such as terminal illness. He also notes that more representative studies using randomized samples indicate that less than half of patients are interested in discussing their spiritual concerns with clinicians.[23] At minimum, Sloan's objections give reasons for advocates of spiritual assessment to be more careful in their use of studies to make claims about patient interest.

If the interest were greater and reliably reported, would this justify greater involvement in the patient's spiritual life? Sloan argues that this does not necessarily follow. He compares patient requests for a spiritual discussion to patient requests to end chemotherapy prematurely.[24] These kinds of request according to Sloan create a conflict between the clinician's goal of beneficence and the obligation to respect the autonomy of the patient. His analogy is not appropriate even if one avows that discussions of spirituality *may* be harmful to patients, as Sloan does. Demonstrating that discontinuing chemotherapy early harms a cancer patient is a more

---

[20] Reis and others: 42.
[21] Ibid.
[22] Ibid.
[23] Sloan, 237–38.
[24] Ibid., 239.

straightforward process than demonstrating that talking about spirituality harms the patient. If a strong argument can be made that spiritual assessment is harmful to patients, then a patient's interest in it would clearly not justify its undertaking. Sloan makes the case that spiritual assessment may cause harm in several ways. First, patients who are told that a religious life can lead to better health are implicitly being told that a failure in health indicates a shortcoming in religious life. This message supports what Sloan calls the "just-world hypothesis," the belief that the occurrence of events, e.g. change in health status, can be reconciled to various modes of justice. Bad things do not happen to good people. Endorsing such a hypothesis generates additional guilt when illness inevitably strikes.[25] A second harm proposed by Sloan involves the encouragement of foregoing or discontinuing vital medical care for religious reasons. He cites the case of Chad Green, a 2-year-old boy with leukemia, whose parent substituted dietary supplements for chemotherapy. They defended these actions by saying that if Chad died it was God's will. His parents ignored court orders to treat their son and Chad died at the age of four. This example demonstrates a harm, but it is based on the tenuous assumption that spiritual assessment would *encourage* rather than *challenge* this kind of reasoning in patients. This weakness in the argument raises the issue of competency.

Is spiritual assessment a task that can be done well by health care professionals? Or is it highly susceptible to abuse? Sloan proposes that the authority attributed to physicians by patients because of their *medical expertise* makes discussion of religion with patients vulnerable to misinterpretation, manipulation and coercion. If physicians take a history that is dedicated to spirituality, then patients who are not spiritual or religious could misinterpret this prioritization as implying that spirituality has an important and special relation to health. Sloan also notes that HCPs who do not proselytize are still in the position to subtly influence, manipulate or coerce. This exertion of authority can be used to promote or discourage the use of religious or spiritual resources by patients. Many of the protections that have been achieved to protect patient autonomy are undermined, Sloan asserts, when spirituality becomes part of the physician's responsibility. "When physicians take on the work of the clergy, they become both bad clergy and bad doctors."[26] The concerns around competencies and abuse generate the best reasons against formal spiritual assessment especially when Koenig openly accepts that a benefit of spiritual assessment is to influence patient compliance and Reis endorses its influential role in decision making. I will defend this claim in the next section.

If patients were not interested in spiritual assessment but it was demonstrated as beneficial to their health or their decision making, then does this justify raising spirituality in medical encounters? Reis and Koenig make claims to the effect that spiritual assessment can improve health outcomes. Health outcomes depend on many factors. Reis suggests that spirituality has a positive association with coping, self-esteem and support systems. Sloan concentrates on the methodological weaknesses of studies that report correlation between religious involvement and health

---

[25] Ibid., 187.
[26] Ibid., 206.

outcomes. Studies that do apply rigorous controls and thorough statistical analysis to establish a correlation are still left with the questions of causality. For example, what causal factors can be theorized about the statistical association between church attendance and reduced mortality?[27] Sloan does not direct his attention to claims about spirituality and the psychosocial dimension of health outcomes, e.g. that spirituality is positively associated with health behaviors such as coping with pain and chronic illness.[28] Presumably, if these studies were methodologically sound, then he would raise the concern that some patients would discontinue medical treatment if they were told that spirituality is efficacious in dealing with pain.[29] This objection depends on the quality of the spiritual assessment and the ability of the HCP to show the patient that both are needed. Nonetheless, Sloan raises important objections to the whole enterprise of spiritual assessment including its unintended outcomes.

Explorations of Spiritual Assessment in Genetic Counseling

In this section, I explicate and analyze research on two assessment tools that attend to the spiritual life of patients undergoing genetic counseling. The HOPE[30] approach (see Appendix) is exclusively focused on spiritual assessment and leaves open the question of whether this kind of content should be integrated into psychosocial assessments. The second intervention, named the Colored Eco-Genetic Relationship Map (CEGRM) assesses a number of psychosocial factors including spirituality and religion. A detailed look at these approaches is necessary to see what exactly spiritual assessment might entail either as a stand alone intervention or as part of a psychosocial pedigree. In my evaluation of these approaches, I suggest several reasons why genetic counselors should not incorporate spiritual assessment as a standard part of their practice.

---

[27] Ibid., 142.

[28] For sample of this literature see: A. Bussing, T. Ostermann, and H. G. Koenig, "Relevance of Religion and Spirituality in German Patients with Chronic Diseases," *Int J Psychiatry Med* 37, no. 1 (2007), M. O. Harrison and others, "Religiosity/Spirituality and Pain in Patients with Sickle Cell Disease," *J Nerv Ment Dis* 193, no. 4 (2005), H. G. Koenig, "Religion and Medicine Iv: Religion, Physical Health, and Clinical Implications," *Int J Psychiatry Med* 31, no. 3 (2001), A. B. Wachholtz, M. J. Pearce, and H. Koenig, "Exploring the Relationship between Spirituality, Coping, and Pain," *J Behav Med* 30, no. 4 (2007).

[29] Sloan cites several studies that indicate low psychiatric uptake of patients who think prayer worked for mental health conditions. Sloan, 187–89.

[30] Reis and others: 45.

HOPE Approach

In 2007, Reis et al. published findings from a survey that elicited genetic counselors' attitudes toward and practices of spiritual assessment in genetic counseling. It also reported responses to a possible instrument for facilitating spiritual assessment in genetic counseling called the HOPE Approach. The researchers acknowledge that the survey had a low response rate and hesitate to generalize to the larger population of genetic counselors. In terms of undertaking spiritual assessment in current practice, 60 % of the 127 respondents had conducted a spiritual assessment in the past year and within this subset the mean frequency of performing such an assessment was 20 % of the cases with only 8.5 % conducting in more than half the cases. Those who had performed spiritual assessments in the past year acknowledged being comfortable doing them and indicated their relevance at a significantly higher level than those counselors who had not done one in the past year.

Several tentative conclusions are drawn from these results. In terms of the prevalence of spiritual assessment, the findings permit the conclusion that this kind of exchange does not occur frequently even with counselors who have been willing to perform such an assessment in the last year. Three kinds of circumstances can be identified as prompting spiritual assessment. The first and most common (76.4 %) reason for doing a spiritual assessment is a patient raises the spiritual or religious concerns. This signal overrides the presumption that all counselees are uncomfortable talking about religion. A second kind of circumstance that motivates spiritual assessments is a function of sessions "that involve termination" (41.7 %). Further delineation is not provided to distinguish between discussions of the termination option at an early stage before an invasive test has been performed or at a later stage when the diagnosis is known and an actual decision about termination has to be made. The third set of circumstances where spirituality is addressed by the genetic counselor involves end-of-life issues (29.9 %).[31] In reference to the last two circumstances, Reis et al. surmise that genetic counselors might know more about spiritual issues in these circumstances and consequently are more willing to either elicit or respond to religious concern in these areas. The identification of actual reasons that motivate genetic counselors to engage in spiritual assessment inform the normative question of whether spiritual assessment should be a standard feature of all genetic counseling sessions. Along the same lines, the reasons that they perceive as preventing them from engaging in this domain are also relevant (Table 5.1).

Although empirical findings do not settle the question, the survey does identify reasons that are candidates for more general arguments about the role spiritual assessment should play. Below is a replication of a table in the article[32]:

> These barriers are treated as obstacles to be overcome rather than to be accepted as will be shown below. Counselors had five areas of concern: 1) insufficient skill 2) insufficient time 3) client discomfort 4) counselor discomfort 5) low relevance. A brief review of these will fill out the picture of why the respondents have not undertaken spiritual assessment.

---

[31] Ibid., 44.
[32] Ibid., 45.

Table 5.1 Barriers to spiritual assessment in genetic counseling

Total sample: N = 127	N	%
Survey supplied barriers (Close-ended)		
There is not enough time in the session	58	45.7
I think the client would be uncomfortable discussing spirituality	35	27.6
I do not know how to assess spirituality	22	17.3
I would not know what to do with the information	19	15
I do not think that the client's spirituality is important	14	11
I am uncomfortable discussing spirituality	9	7.1
I am not a religious/spiritual person	8	6.3
My own religious beliefs might conflict with those of the client	6	4.7
My spirituality might conflict with my client's	6	4.7
Spiritual assessment is the job of chaplains and clergy members	2	1.6
Respondent supplied barriers (Open-ended)		
Client did not bring up spirituality	19	14.9
Spirituality was not relevant to the session	16	12.6
I do not think assessment is necessary in basic GC sessions	10	7.9
Spirituality did not seem to be important to the client	8	6.3
The client resisted discussing the topic	2	1.6
Spiritual assessment is not my role	1	.8
The physician I work with is uncomfortable with the topic	1	.8
Spirituality is not assessed in follow-up sessions	1	.8
Respondents were allowed to select more than one option.		

A significant set of obstacles revolves around the perception of having insufficient skills to address spirituality. Genetic counselors acknowledge a lack of assessment competencies and an understanding of what to do with such assessments. The authors' response is: More education would make spiritual assessment more likely. How likely depends on several other factors. If the number one barrier given is insufficient *time*, then sufficient skill would not necessarily change this constraint. Reis comments: "While there is little that can be done directly to reduce this barrier, it seems likely that willingness to make time for spiritual assessment will increase as the perceived value of such assessment increases."[33] The ability to 'make time' implies that the perceived time constraint is not real or that time spent covering other topics can be reallocated to spiritual topics. In my limited observations of genetic counseling sessions, the scarcity of time was observable in the challenges of

---

[33] Ibid., 47.

coordinating the activities of genetic counselors, sonographers and OB/GYNs. Third, over a quarter of the genetic counselors assume that *counselees* will be uncomfortable discussing spirituality. The authors highlight that this perception is incompatible with at least three studies that indicate patients are interested in discussing spirituality with health care providers. This counterevidence does not mean that the genetic counselors are wrong to make such attributions in any given case but only that they may be overestimating the level of discomfort counselees have. Fourth, the authors report that about one third of the respondents are uncomfortable with spiritual assessment. Citing a separate study, they identify a list of possible sources of discomfort:

> Discomfort with spiritual assessment may stem from a fear that just as psychosocial counseling follows from psychosocial assessment, spiritual assessment will create a need for spiritual counseling...In addition, counselors may feel uncomfortable with the unresolved dilemmas surrounding spiritual assessment in health care including the subjective nature of spirituality, the appropriate roles of members of the health care team in providing spiritual assessment and care and uncertainty regarding the careful balancing of the needs for confidentiality and documentation of the information gained through spiritual assessment.[34]

No challenges or possible solutions are offered to this important set of concerns that do not appear easily resolved. A likely proposal would again look to the training and education of genetic counselors to mitigate some of these concerns. Fifth and finally, another third of the respondents indicated that spirituality is not relevant to genetic counseling. The reasons why genetic counselors consider spirituality to have low relevance is not made explicit in the study but counselors who expressed low comfort also indicated low relevance. Thus, the concerns above might also lead to the conclusion that spirituality has low relevance in genetic counseling. These barriers, despite the proposals to overcome them, complicate the question of whether spiritual assessment should be undertaken in genetic counseling. In their attempt to understand the different stances toward spiritual assessment, have Reis et al. undermined their own interests in promoting spiritual assessment?

A brief discussion of the responses to the HOPE tool for spiritual assessment will provide elaboration of what is meant by spirituality and an indication of different responses to its various domains. Spiritual assessment can take the form of informal discussions with patients or it can be a formal process of answering questions. In this section of the survey, Reis seeks to identify responses of relevance and comfort to a formal spiritual assessment tool called the HOPE questions (See Appendix). Anandarahajah and Hight, both physicians, created an assessment tool called the HOPE questions that could be used in the practices of or training of physicians.[35] Each letter stands for a group of questions focused around specific areas of spirituality. 'H' questions elicit what are the patients' sources of "hope, meaning, comfort, strength, peace, love and connection" as a non-threatening way of assessing the resources the patient possesses to respond to difficult circumstances. If a patient indicates that spiritual or religious beliefs do provide them comfort, then the

---

[34] Ibid.

[35] Anandarajah and Hight.

practitioner should ask questions under 'O' and 'P.' The 'O' questions concern the patient's status as a participant in an organized religion. The inquiry probes the level of participation and raises evaluative questions about how helpful religion is to them. Having addressed spiritual expression through organized religion, 'P' questions explore patients' personal spiritual practices that are independent of their religious community. Patients are asked to identify and evaluate their spiritual practices. There is also a question about whether the patient believes in God and if she does, how would they describe their relationship with God. The final set of question look to "E"-ffects of spiritual practices and beliefs on medical care and end-of-life issues. The questions address both the ways health circumstances affect spiritual activity and the way that spiritual needs can affect the delivery of health care. The answers to these last questions have the most observable affect on the actions of health care providers, e.g., withholding blood products, requesting a chaplain, or praying with a patient. Anandarajah and Hight locate these question within a large conceptual framework that defines and argues for spiritual assessments.

The genetic counselors were asked to respond to a randomized list of HOPE questions in two ways: (1) indicate, yes or no, whether the question was relevant in some circumstances (2) express how comfortable they would be asking the question on a Likert scale. The results indicated that genetic counselors were increasingly comfortable with the questions they found most relevant. The 'H' questions were relevant[36] to 93.4 % of respondents and 53.4 % said they would be comfortable with asking all the questions in this section. The 'E' questions ranked second in relevance (86 %) and comfort (27.3 %). The 'O 'questions were a distant third in terms of relevance (49.6 %) but similar in comfort level (24.8 %) with 'E' questions. Finally, the 'P' set had lowest relevance ranking (31.4 %) and a significantly lower comfort rating (5 %).

The differences in relevance and comfort level between question types is striking, and some interpretation of these gaps is offered. The high score for the 'H' question is explained by two factors. First, this kind of content may already be addressed in most genetic counseling sessions and second, the way these questions are worded makes them indirectly spiritual according to the authors. The 'E' questions possibly received higher relevance and comfort scores because genetic circumstances that bring specific types of medical care or raise end-of-life issues have been acknowledged in psychosocial or culturally-sensitive training resources.[37] Concerns about intruding into patients' private lives might explain why genetic counselors thought questions about organized religion were not relevant except in cases where genetic disorders have a high correlation with certain religious groups, i.e. Tay-Sachs and Ashkenazi Jews. The questions about personal spirituality may have received the lowest relevance and comfort scores because they "address spirituality most directly."[38] Alongside a respect for privacy, this inquiry into spirituality proper may result in a presumption that the counselee would be uncomfortable talk-

---

[36] This percentage is based on the amount of counselors who found at least three out of four questions relevant in a given section.

[37] Reis and others, 48.

[38] Ibid.

ing about such personal beliefs and practices. The reservations that are associated with 'O' and 'P' questions combined with the list of barriers cited above might explain why genetic counselors were not 'highly likely' to utilize any of the questions from the HOPE approach.

Although conclusions drawn from the study cannot be generalized to the larger population of genetic counselors, the survey provides a provisional picture of current practice of and attitudes about spiritual assessment. The authors of this study – along with the developers of the HOPE tool – have reasons for advocating spiritual assessment that have been stated above in reference to findings about patient interest and health outcomes. Although the studies in these areas fail to produce anything close to a consensus, Reis interprets the trajectory of these studies as revealing a widespread interest of patients to discuss religion with their health care provider.

CEGRM

Whereas spirituality and religion play a central role in the HOPE approach, they are late additions in the research concerning the use of a psychosocial assessment tool called the Colored Eco-Genetic Relationship Map (CEGRM). In 2001, Kenen and Peters introduced the CEGRM as a "conceptual approach and tool for presenting information about family and nonkin relationships and stories about inherited diseases in a simple understandable form."[39] Combining several assessment tools into one (genetic pedigree, genogram, ecomap), they intended initially to use the tool for research and hoped that it would eventually become a useful mechanism for conducting clinical interviews and storing genetic and psychosocial information about in standardized way. After undertaking a pilot study of 20 subjects,[40] they published findings from a larger study of 150 women who were members of families associated with hereditary breast and ovarian cancer. In this most recent study they expanded the tool to include the domains of spirituality and religion.

Referring to this domain as "Religious/Spiritual Exchanges," the researchers gathered this information from 35 of the 150 women who participated in the initial interviews. They describe the procedure as follows:

> During the portion of the CEGRM that focused on religious/spiritual support participants were asked whether or not they felt that religion and/or spirituality were an important part of their social world and whether there were any individuals with whom they felt spiritually connected.[41]

---

[39] R. Kenen and J. Peters, "The Colored, Eco-Genetic Relationship Map (Cegrm): A Conceptual Approach and Tool for Genetic Counseling Research," *Journal of Genetic Counseling* 10, no. 4 (2001): 289.

[40] J. A. Peters and others, "Exploratory Study of the Feasibility and Utility of the Colored Eco-Genetic Relationship Map (Cegrm) in Women at High Genetic Risk of Developing Breast Cancer," *Am J Med Genet A* 130, no. 3 (2004).

[41] J. A. Peters and others, "Evolution of the Colored Eco-Genetic Relationship Map (Cegrm) for Assessing Social Functioning in Women in Hereditary Breast-Ovarian (Hboc) Families," *J Genet Couns* 15, no. 6 (2006): 482.

They elicited these feeling with prompts like:

> Some people talk about a religious sort of connection, such as knowledge of a shared faith, attending services together, or praying with or for each other. Others talk about a less definable more ethereal kind of connection or closeness with other beings or even finding a peaceful place within oneself. Are any of these important to you?[42]

Although no formal definition of either spirituality or religion is offered to underwrite this prompt, it appears that they – similar to the HOPE approach – were proposing an expansive definition in hopes of capturing the full range of attitudes about spirituality. Their findings reveal some success on this front having received answers ranging from participation in traditional prayer groups to meaningful connection with pets.[43]

In analyzing the addition of a spiritual/religious component, two results are mentioned along with an acknowledgment of the complexity of undertaking this kind of assessment. By adding this domain, the researchers benefited by getting a more comprehensive picture of the psychosocial situation. How exactly it improved their understanding is not specified. One can assume that they obtained more information and that they had new insights about the different kinds of social exchanges that define the participant's world. The benefit to the participants is conferred in the feeling of being "understood in a holistic way that promotes healing."[44] They justify this ascription by endorsing the idea "as defined by Egnew to relate to the personal subjective experience involving reconciliation of the meaning an individual ascribes to distressing events with her perception of wholeness as a person. "[45] Without elaborating at this juncture, I want to point out the link between this effect of the CEGRM and the therapeutic effects of communication introduced in Chap. 2. Carl Roger's vision of the healing powers of empathic understanding is transposed to a psychosocial pedigree that can provide holistic understanding that promotes healing. Finally, the researchers acknowledge that spiritual assessment is a complex activity that entails a large subset of domains to be explored. They cite several recent studies and allude to a growing awareness in the counseling profession (and also genetic counseling) about the importance of these domains for understanding the client.[46]

The CEGRM is an important model because it incorporates spiritual assessment within a comprehensive conceptualization of the psychosocial profile of patients. Reis et al. identified the importance of deciding whether spiritual assessment needed to be treated separately or as part of the psychosocial assessments that many genetic counselors have been trained to perform. The CEGRM is one way of working out

---

[42] Ibid., 480.

[43] Ibid.

[44] Ibid., 485.

[45] Ibid.

[46] Two studies they cite are: M. Stefanek, P. G. McDonald, and S. A. Hess, "Religion, Spirituality and Cancer: Current Status and Methodological Challenges," *Psychooncology* 14, no. 6 (2005).; E. A. Rippentrop and others, "The Relationship between Religion/Spirituality and Physical Health, Mental Health, and Pain in a Chronic Pain Population," *Pain* 116, no. 3 (2005).

that relation and could become part of the training process in genetic counseling.[47] Spirituality and religion are interpreted as important connections within a larger web of social relations that have a bearing on health concerns. By addressing this complex domain, both the genetic counselor and the patient appear to receive benefits from this deeper understanding. Whether patients or clients understand their own spiritual life as only one component of their social identity is an important question given that religious beliefs are often used to interpret the whole of one's identity.

Evaluation of Harms and Benefits

In this section, I evaluate the prospect of whether spiritual assessment should become a standard component of genetic counseling. To adopt spiritual assessment means in this context applying an instrument like the HOPE approach in all initial and if necessary subsequent genetic counseling sessions. Both potential benefits and harms receive examination. First, what are the potential harms of conducting a formal assessment? If there are likely and significant harms, then spiritual assessment should not be undertaken. If the harms are unlikely and insignificant, i.e. closer to inconvenience, then they must be weighed against the potential benefits. If benefits are likely and significant and harms unlikely and insignificant, then the feasibility becomes an important question. I conclude that there are some significant harms that are likely, but their severity does not preclude a weighing of possible benefits. After evaluation, the benefits that Reis claims would be conferred look less promising. Thus, increased time allowance and high levels of competence, which would in themselves be very difficult to achieve, would not change the conclusion that formal spiritual assessment is not appropriate in genetic counseling.

Sloan suggests above that formal spiritual assessment may cause several harms and similar ones are identified in the Reis study. Three will be evaluated here: (1) lack of relevance, (2) patient privacy, and (3) misunderstanding and manipulation of patients. In the Reis survey, the top two responses to open-ended questions about barriers pertain to the issue of relevance. The respondents commented that "clients did not bring up spirituality" and that "spirituality was not relevant to the session." Sloan would use these findings to bolster his argument that separate spiritual histories give disproportionate attention to one sphere of possible concern. He does not believe that spirituality or religion deserve to be separated from the other activities, e.g. sports, and relationships, e.g. family, that are important in patients' lives and that have associations with health status. This argument is especially compelling when considering whether to institute *routine* spiritual assessments such as the HOPE approach in genetic counseling. Genetic counselors cannot assume without prior knowledge of the patient that spirituality is a relevant concern. The question is

---

[47] Peters and others, "Evolution of the Colored Eco-Genetic Relationship Map (Cegrm) for Assessing Social Functioning in Women in Hereditary Breast-Ovarian (Hboc) Families," 487.

whether it is harmful to bring it up without entitlement to this assumption. Leaving the discomfort of declining to talk about such matters aside, patients may be irritated that irrelevant questions were asked and may feel harmed if time does not permit them to address concerns they consider important. Sloan cites a study where patient interest in discussing spirituality plunged (to 10 %) if discussing it meant that other medical concerns would not be addressed.[48] Since lack of time is the number one barrier identified by genetic counselors in the Reis study, spiritual assessments such as the HOPE approach, which involves several questions, may cause harm to those patients who do not feel the important issues have been given enough time. The CEGRM, which has only one open-ended question about spirituality, requires less time and is less likely to harm patients who do not consider spirituality important.

If these instruments make some counselees irritated or uncomfortable, then one reason for this discomfort might be that patient privacy is not being respected. One consequence of the U.S.'s commitment to separation of church and state is that many citizens in the U.S. consider their religious beliefs to be private or intensely personal. For different reasons, many people also consider their health status to be a private matter. The reason that genetic counselors are authorized to routinely discuss private issues concerning genetic status is they have specialized knowledge in this area. Since they do not have authority in matters spiritual, it raises the question of whether preemptive questioning is an invasion of patient privacy. The legal answer is no. Initiating talk of spirituality does not infringe on legal privacy protections unless this information is shared inappropriately by the HCP. Is this a professional invasion of privacy? The answer here is yes with some qualifications. By entering into a relationship with an HCP, a patient confers trust that the HCP will only assess that which is relevant to help the patient and that which is not relevant will be left undisturbed. The question is how to address those areas that *might* be helpful. From the perspective of privacy, HCPs should assess these *potentially* helpful areas in the least invasive way possible. The HOPE tool does not meet this criterion. Sloan is right to argue that setting spirituality aside as a special topic "implies a degree of importance based on the physician's values, not necessarily the patient's."[49] According to the results of the CEGRM, patients benefited from having their total perspective taken into account. Despite this evidence, I do not think that it warrants inserting into a conversation an assessment tool like the HOPE instrument because it probes an area that the patient potentially considers too private or personal. The HOPE approach creates more risk than the CEGRM because the latter is much less invasive. And in terms of privacy, this is an important difference. If the CEGRM were to minimize its invasiveness, the HCP would invite the patient to discuss his or her personal concerns and list examples to suggest areas that might be of concern. Spiritual or religious concerns could be one example in the list but should not be set out as special.

---

[48] Sloan, 238.
[49] Ibid., 193.

The final harm to consider pertains to the susceptibility of spiritual assessment to misunderstandings and manipulation. The emphasis genetic counseling often has on reproductive decisions makes this set of concerns especially relevant. The possibility of these harms is related to the feasibility of standardizing the competencies and commitments the HCP has in this kind of assessment. The multiple valences of spiritual and religious meanings is not unique to these vocabularies and practices but assessing spirituality places attention on a set of meanings that could be difficult to navigate or negotiate depending on the patient's commitments and the genetic counselor's competencies. If they were unable to coordinate their perspectives, then this could lead to misunderstandings that affect decision making. For example, in Debbie's case, how should the genetic counselor understand her remarks about God's will? Is Debbie still open to termination, or do her religious comments override her earlier comments? If the HCP had initiated this conversation, Debbie's comments might have been different because they were elicited in the context of an assessment. The fluid and malleable status of many patients' religious beliefs – especially the commitments of patients who would only talk about spirituality in response to the HCP's spiritual assessment – could lead to important misunderstandings. In circumstances where patients are trying to sort out their perspective, the formal assessment of their spiritual resources may actually inhibit this process if patients do not usually consider such concepts in decision making.

Although religious/spiritual beliefs are sometimes characterized as dogmatic and by implication stable, they can also be loosely grasped and seldom used. In the presence of an authority like an HCP who presumes competence to bring up spirituality, the patient may be vulnerable to manipulation. Manipulation in this case would involve exploiting the patient's stances on spirituality to achieve a specific outcome. For example, as Koenig points out spiritual assessment can influence patient compliance. If a genetic counselor has reservations about amniocentesis and termination, then the spiritual assessment could easily be used to manipulate a conflicted client away from this procedure. Six genetic counselors in the Reis survey acknowledge that their spiritual beliefs might conflict with the patients'. One can speculate that if spiritual assessment became standard practice, this number would rise. A stronger claim is that a standardized practice of spiritual assessment is inherently a manipulation by the group of HCPs who institute it. By making it standard, they are manipulating what Mary Douglas has called "the thought style" of patients.[50] On this view, a spiritual assessment manipulates patients to think in certain categories. Although I do not think this is the motivation of those who promote spiritual assessment, it may be an unintended harm of this practice. Of the three set of harms evaluated, I think manipulation is the most significant because of its potential impact on decision making.

At the beginning of this section, two main reasons were cited as motivating the interest in formally assessing spirituality. Patients are interested in having these

---

[50] Mary Douglas, *Risk and Blame: Essays in Cultural Theory* (London; New York: Routledge, 1992), 211. Douglas is using the term to refer to the thought style of a culture but it is appropriate for individuals as well.

concerns addressed and acknowledging spirituality improves health outcomes. In several studies, a significant percentage of patients indicated that spiritual concerns were important to them and that they were interested in physicians exploring their spiritual concerns.[51] Researchers in genetic counseling have cited these studies as reasons to consider instituting formal spiritual assessments. One benefit this would confer is the showing of respect for what is important to the patient. Sloan's objection to the premise that studies have shown patients are interested do not need to be repeated here. Until rigorous studies show that a clear majority of patients – such studies could have a regional emphasis – want HCPs to raise issues of spirituality, the claim is tenuous that formal spiritual assessment shows respect for the patient preferences.

The second benefit that supports formal spiritual assessment is improved health outcomes. As stated above, Reis concludes in the wake of 20 years of research that "spirituality has positive effect on mental physical and emotional health."[52] Although Sloan's arguments bring any such claim under suspicion, the more important question is whether a formal spiritual assessment actually promotes the positive effect that spirituality can have on health outcomes. Thus, even in light of research that patients who are 'spiritually active' have better health, it remains to be demonstrated that a formal intervention would actually augment the association. For example, bringing up spirituality routinely in the context of prenatal counseling could consistently have a negative effect on the psychosocial states of patients who otherwise would not have used spiritual or religious terms to think about their situation. When the HCP utters words like "religion" or "spirituality," some patients will interpret the meaning from their own perspective. A patient's association with these terms is unpredictable. Other patients may wonder what the HCP means by them, whether there is an agenda behind them. This uncertainty may bring anxiety that is caused specifically by the introduction of these terms in a series of questions. Formal spiritual assessment may actually undermine the positive association between spirituality and health outcomes in these circumstances. The inconclusive relationship between *spiritual assessment* and positive health outcomes weakens the claim that a benefit would actually be conferred by routinizing spiritual assessments.

The final benefit that spiritual assessment could confer is improved patient decision making. Reis points out correctly that "religious and spiritual beliefs can profoundly influence medical decision-making."[53] She supports this claim with two studies. The first reports that 45 % of adult patients in a pulmonary care center, if they became gravely ill, would use religious or spiritual commitments in their decision making. The details of this study lack specific application to the context of genetic counseling and raise the question of relevance. The second study directly involves genetic counseling circumstances. Patients at high-risk for breast cancer received pretest genetic counseling for BRCA1/2 as part of study to see whether

---

[51] See studies cited above.
[52] Reis and others, 42.
[53] Ibid.

psychosocial factors and spiritual faith had influence on test use.[54] Women who had a low risk level of recurrence were less likely to be tested if they had strong level of faith than if they had low levels of faith. The study indicated no difference of test use between women at high-risk level. What is not explicit is how this empirical study informs the normative question. Does the correlation between faith level and test uptake support or undermine the position that genetic counselors should undertake spiritual assessment? Advocates might say that it is important for patients to be aware of their spiritual status because it may be relevant for their decision making regarding genetic testing. This much can be endorsed. A standard spiritual assessment would provide patients with the opportunity to achieve this awareness of "faith level" in relation to the specifics of the genetic circumstances. The concern is that the preemptive nature of the questioning would alter the moral thought style of many patients, even the individuals who are self-consciously religious. Genetic counselors already have a large influence on the shape of the conversation. To add a spiritual assessment to the pedigree and other intake procedures gives the genetic counselor control over one more vocabulary. I have tried to show that the potential harms are significant and the benefits are tenuous.

The responsibility approach proposed below suggests that genetic counselors should be responsive to the thought style that emerges in the dialogical process. In Debbie's case, religious concerns were raised toward the end of the session. The focus of the analysis will be on the way the three models purport to handle spiritual and religious matters.

Spiritual Assessment and Debbie's Case

The rest of the chapter entails a brief exploration of (1) the three models' stances on spiritual assessment and (2) the three models' approaches to the spiritual and religious concerns in reference to a Debbie's case. As the previous chapters have shown, genetic counselors have several models available to guide their communication strategies and tactics. The teaching model undertakes the primary goal of educating the client by transmitting the genetic information with clarity and at a level appropriate to a patient's apparent ability to understand. The psychotherapeutic model has multiple goals. The genetic counselor seeks to understand the client, enhance the client's confidence in decision making, promote her autonomy, resolve psychological distress and help her find specific solutions to the problems at hand. The responsibility models aims to coordinate the meanings of genetic information across diverse perspectives and facilitate responsible decision making. I suggest how the three different models approach the case and conclude that the responsibility model provides the most adequate resources for understanding how the HCP could help Debbie interpret the genetic information.

---

[54] M. D. Schwartz and others, "Spiritual Faith and Genetic Testing Decisions among High-Risk Breast Cancer Probands," *Cancer Epidemiol Biomarkers Prev* 9, no. 4 (2000).

Before turning to the models, some general insights and observations can be made about the case in relation to the issue of spirituality. The definition of spirituality proposed above provides the genetic counselor with a model to identify several spiritualities within Debbie's complex stance. Patients often trust more than one kind of spirituality when limiting conditions such as uncertainty about health status arise. That Debbie took time to come to the medical center implies that she is a participant in what Sheridan calls axial spirituality, the subculture that emphasizes *human* resources for transforming our limiting conditions. The biomedical tradition is thus understood within a much broader kind of spirituality that emphasizes what is sometimes called the human spirit. As the session unfolds, Debbie's comments indicate that she also participates in a religious practice concerned with the will of God. Her intentional stance towards the various outcomes reflects a trajectory of meaning that has at least in part been fed by Christianity. In Sheridan's model, religion is continuous with axial and plenum spirituality and also distinct because of its concern with the assistance of a Transcendent Other. Although it is tempting to ascribe to Debbie a contradictory stance that trusts biomedicine and religion, Sheridan's taxonomy allows us to see Debbie as a bricoleur[55] of spiritualities.

With this framework, specific features of the case become more coherent. First, Debbie's use of religious terms occurs as the uncertainty of her situation takes hold. The genetic counselor has attempted to transform the uncertainty by offering probabilities[56] and possible outcomes. Debbie's turn to religion is not a way of dismissing the genetic counselor. It is a signal that the uncertainty surrounding prenatal diagnosis has not been completely transformed for her by probabilistic statements. Her religious beliefs are an additional and available cultural resource to address her uncertainty but such resources are not available like tools in a hardware store. They are available as intentional stances that functions as established nodes within the pulsing network of a forged identity. Thus, what seems like a dramatic shift in content to an observer makes sense as a complex perspective calling on available resources to address uncertainty. Finally, the genetic counselor's response to the religious utterance is to give Debbie and her spouse some privacy to deliberate. This move does not necessarily reflect discomfort with religion. More likely it reflects an enactment of nondirectiveness supported by the teaching model.

---

[55] For appropriations of Claude Levi Strauss's notion of 'bricoleur that have influenced me, see Jeffrey Stout, *Ethics after Babel: The Languages of Morals and Their Discontents* (Boston: Beacon Press, 1988), 74. and Churchill and Schenck: 401.

[56] Sheridan notes that culture is a paradox that generates and transforms certain human predicaments. In this case, generating a risk status creates a new predicament of understanding what that means in practical circumstances.

Teaching Model

In the previous chapter, the features of Debbie's case were interpreted through the categories of directive and nondirective counseling. It was established that the genetic counselor's moves in Debbie's case are best understood as being guided by the teaching model. The genetic counselor achieves her goal by doing three things (1) offering a risk assessment (2) explaining amniocentesis and its risks (3) laying out the practical alternatives and outcomes. When these tasks are complete, the genetic counselor should not interfere with the deliberation process. Ascribing to the teaching model has several implications for how spirituality and religion might be addressed.

A formal spiritual assessment as defined by the HOPE and CEGRM tools would not be permissible under the teaching model. The intervention is incompatible with the model for several reasons. First, the spiritual and religious content is not information that the genetic counselor is competent to send or receive. By initiating conversation about religion, the genetic counselor would be undertaking authority in an area in which he or she cannot take responsibility. Some genetic counselors may feel uncomfortable raising spiritual concerns because the need for spiritual or pastoral counseling might arise. This discomfort is an outgrowth of the lack of authority. Second, raising spiritual issues could send an implicit message to the counselee that he or she should be using these kinds of resources to understand and make decisions. Such a message risks being directive by the teaching model's standards because it directs the counselee towards a certain kind of evaluative framework. Finally, religious and spiritual content is not the kind of information that is easily transferred between people. It is a subjective form of belief that is not easily standardized across perspectives and as a result is highly susceptible to confusion or manipulation.

Under the teaching model, a genetic counselor can correct misconceptions of genetic information. Presumably, this includes misconceptions rooted in religious commitments. In Debbie's case, she states that God's will would be manifested in two different outcomes: the birth of a child with Down syndrome and a miscarriage caused by amniocentesis. Is this a misconception of the situation? Should the genetic counselor correct it? It would be difficult to classify Debbie's comments as misconceptions if guided by the teaching model. Debbie's religious utterances are on this view highly subjective and not easily transmitted to the HCP. The vocabularies of genetics and religion are incommensurate from an educational standpoint. The genetic counselor who follows the teaching model would not be entitled to challenge or correct this statement because it is not in the HCP's realm of authority. Efforts to help Debbie elaborate the concept of God's will are also outside the competencies of the genetic counselor who is trained to transfer objective, scientific information. In light of these constraints, the offer of privacy is a reasonable move for the genetic counselor to make.

The psychotherapeutic model challenges the teaching models on two fronts. First, the teaching model's stance toward religion is a symptom of a larger problem of not trying to understand the patient. It does not take seriously enough the ability of the counselor to bracket his or her commitments and achieve an empathic, if provisional, understanding of the client perspective. By narrowing the authority of the professional to only matters concerning scientific knowledge and options, the teaching model isolates the patient. The second set of problems with the teaching model's approach to spirituality is that it ignores a valuable resource for promoting the patient's autonomy. In Debbie's case, reference to God's will could have been construed as empowering to make a decision with the confidence that she could handle any outcome because God would be present. The teaching model misses this opportunity to bolster the patient's confidence and resolve the psychological anxiety of practical uncertainty.

The teaching model makes two critical errors in its approach to religion according the responsibility model. First, it fails to appreciate the holistic structure of meaning and as a consequence has the false assumption that the meaning of the genetic information can be delivered as a tightly contained semantic package that is then applied by the counselee. This misunderstanding can be accounted for in the atomistic semantics of the technical vision of communication. This view of meaning underestimates the role of dialogical processes in actually grasping the meaning of the information. This leads to the second error. The teaching model fails to see the responsibilities of coordinating the meaning of genetic information from both the HCP's and Debbie's perspective. The teaching model is right to recognize that inferential restraint and humility is a necessary part of negotiating meaning but is wrong to assume that Debbie's beliefs are radically different and disconnected from the genetic information. The two perspectives together can expand the meaning of God's will in helpful ways. For example, the genetic counselor could ask Debbie what it would mean for God to will that she have a disabled child or miscarriage. This prompt would give Debbie the opportunity to trace out the consequences of her statements, and it would require no theological but only dialogical competence on the counselor's part.

Psychotherapeutic Model

Of the three models, the psychotherapeutic model would be the most likely to utilize a spiritual assessment tool. Reis and Peter's interest in spirituality is rooted in a broader concern for addressing the psychosocial needs of the patient. The HOPE approach does provide a thorough instrument for understanding the patient and would signal to them that their perspective is important. The CEGRM is a psychosocial assessment tool that would be integrated into the taking of a patient's pedigree. It is more compatible with the psychotherapeutic model than the HOPE approach because it treats spirituality as part of a larger psychosocial picture. Since CEGRM was conceived as being more useful with families that face specific heritable diseases, whether it should be applied in prenatal counseling remains an important question.

Other advocates of the psychotherapeutic model support genetic counselors' initiating talk of religion. John Weil proposes that genetic counselors should raise the issue of religion "since the counselee may not consider these topics to be relevant, appropriate, or of interest to the counselor."[57] He suggests that the counselor provide cues or asks questions more along the line of the CEGRM than the HOPE approach: "'Do you see yourself as a religious or spiritual person?' Or 'Do you have a religious or spiritual community that would be helpful?'"[58] If the patient answers in the affirmative, then the counselor is entitled to direct the patient to think about these religious resources in relation to decision making and as a support mechanism. All of these questions are a way of acknowledging the importance of the patient's perspective. If the counselee is conflicted by tension between personal values and her religion, the genetic counselor, without trying to resolve the conflict, should attempt to empathically understand the torment that the patient experiences. Weil rejects the assumption that an HCP needs to know something about the religion to deal with religious concerns. Instead basic counseling skills are all that is needed to reach the goals of the psychotherapeutic model.

From Kessler's and Resta's opening quote, it appears that they are willing to adopt a robust religious standpoint on behalf of the patient. In a situation different from Debbie's, the commitment to empathic identification entitles them to apply a counseling skill that has substantive theological content, content that they may not actually endorse:

> God sometimes gives people special tasks in life that we need to do without rewards or thanks and yours is to care for Alexis until He's ready to take her. And, the two of you have done a magnificent job in caring and loving her. You may not experience her love now, but a day will come when you will. God will take care of that.[59]

Kessler acknowledges that this statement would be challenged by those who hold that the HCP should dispel false hopes rather than perpetuate them. Kessler defers to the patient's coping system and allows it to determine the meaning of the event. The genetic counselor is empathically adopting the framework of the patient.

All of these stances on spiritual assessment give insight into how Debbie's case might have been handled by a genetic counselor who avows the psychotherapeutic model. Imagine that a version of the CEGRM, which would be customized to prenatal counseling circumstances, is adopted into the psychotherapeutic approach.[60] The CEGRM involves asking the counselee questions about their genetic identity as well as their psychosocial identity. Initially questions would be asked about Debbie's and her family's health status. Subsequently, psychosocial questions would be asked. The emphasis of the psychosocial dimension is the network of exchanges in which the patient is involved. For example, with whom does the patient share

---

[57] Weil, *Psychosocial Genetic Counseling*, 52.
[58] Ibid.
[59] Resta and Kessler, "Commentary on Robin's a Smile, and the Need for Counseling Skills in the Clinic," 437.
[60] The researchers who are developing the CEGRM have never indicated that it would be used in prenatal genetic counseling for advanced maternal age.

information? With whom does she share emotional content? With whom does she share spiritual/religious exchanges? These questions intend to show not only the HCP but also the patient the structure of a support network and the patterns of using it. After gathering this information and making the appropriate inferences, the same probabilities and descriptions would be given. What would happen next is more difficult to predict because the CEGRM changes the interaction presumably in a way that improves the outcome.

This consequence needs to be highlighted because Debbie's comments about God's will and a miraculous gift may not have developed in the same way. This potential shift points to a tension between adopting an empathic stance and adopting a tool like the CEGRM. Whereas empathic identification seeks to bracket the commitments of the HCP and let the patient control the conversation, an assessment tool reinforces the biomedical perspective's propensity for gathering and analyzing information. The prompts cited in the CEGRM study are open enough to be somewhat compatible with empathic goals, whereas the HOPE questions would insert too much structure into the interaction.

Assuming that Debbie did articulate the same concerns after undergoing the CEGRM, the psychotherapeutic model would have several recommendations for a response. These are guided by the goals that were elaborated in Chap. 2. Kessler and Weil would recommend that the HCP promote the perspective of the counselee as a way of building her confidence in decision making. This position would lead to a follow-up question intended to help Debbie find new insights about God's will in her circumstance. Kessler's quotations above suggest that the genetic counselor would even by entitled to speak *as if* he or she held Debbie's same beliefs. Thus a statement such as "Yes, God would bless any outcome" would be permissible. After discussing these religious implications, the genetic counselor should offer Debbie and her spouse privacy only as a last resort. Instead of abandoning the couple, the genetic counselor should help them sort their thoughts in a way that promotes autonomous decision making. Debbie most likely would have reached the same decision. The difference would be that her perspective explicitly received the support of the genetic counselor.

Those who adopt the teaching model would have a couple of objections to the psychotherapeutic approach. One objection relates to the authority of the HCP to initiate discussion about non-genetic matters. Hsia's recommendation is to only ask questions that facilitate better transfer of the genetic information. Kessler's willingness to adopt the perspective of the patient puts the HCP at risk of being held responsible for outcomes the patient later regrets. If the genetic counselor were to implicitly endorse Debbie's religious beliefs, then the counselor would overstep the bounds of relevance and authority. The second objection to initiating talk about spirituality is that it does not serve the goal of educating the client. Instead the learning process might be hindered because of the emotional and metaphysical dust that gets kicked up when discussing spiritual and religious concerns. Weil and Kessler would rejoin that avoiding the emotional and spiritual issues that inevitably arise in cases such as Debbie's has the effect of abandoning the patient or sends the message that these kinds of concerns are not worthy of discussion.

As stated in Chap. 4, the responsibility model and the psychotherapeutic have overlapping stances that are supported by different explanatory strategies. The importance of understanding the patient's perspective is paramount in both models as is supporting the autonomy of the patient. Notwithstanding these common commitments, two problems with the psychotherapeutic approach can be raised from the standpoint of the responsibility model. First, the commitment to empathic identification can undermine important aspect of the dialogical process. The HCP tries not only to suspend his or her perspective for the sake of understanding but attempts to adopt the perspective of the other for the sake of understanding. Kessler's opening quote is potentially harmful not because it makes him complicit, as the teaching model contends, but because it foregoes the dialogical process at a crucial moment. Debbie needs an additional perspective that is honest rather than a doubling of her own perspective. The ghost of the therapeutic vision of communication is seen in the psychotherapeutic model's attempt at empathic identification.

The second problem is related to the first. The psychotherapeutic emphasis on promoting autonomy through empathy results in the near removal of the HCP's perspective in the decision-making process. One consequence of this de-emphasis can be that it actually undermines the goal of promoting patient autonomy. This can become especially problematic in the area of spirituality. The challenge of bringing up spiritual concerns on the psychotherapeutic account is that spiritual concerns can also undermine patient confidence and promote heteronomy or what Karl Barth calls theonomy.[61] Some religious narratives encourage believers to defer all decisions to God, sometimes making prayer the deliberative vehicle and other times waiting for a sign. In these cases, should the genetic counselor try to reframe the situation to promote the client's autonomy, or should he or she respect the patient's obedience to a higher decision-making power? The answer to this question has important normative implications. To reframe the situation is a direct challenge to the patient's beliefs; to support the patient's deference to a sign from a higher power puts the HCP in the difficult situation of promoting a form of heteronomy. Weil would advise the genetic counselor to empathically understand the patient and restrain from challenging or adjudicating her commitments. The responsibility model would also respect a patient's deference to a divine authority but would ideally facilitate an evaluation of this stance through a dialogical process.

Responsibility Model

Of the three of models under consideration, the responsibility model would be less likely to undertake a formal spiritual assessment than the psychotherapuetic model but more likely than the teaching model. Instruments such as the CEGRM and the HOPE approach add additional layers of structure to the dialogical process. In a

---

[61] Karl Barth's notion of theonomy stands as the ground of autonomy and heteronomy. Being bound by God's word (or law) is the very source of the distinction between the latter two terms.

conversation that is already highly constrained by an intake with a pedigree, the addition of a scripted set of questions about spirituality has the unintended effect of perspectival creep. The HCP in an attempt to learn more about the patient may unintentionally lose this opportunity by, so to speak, leading the witness. From the standpoint of the responsibility model, the coordination of perspectives is undermined if one of the perspectives has too much control over the structure and trajectory of the conversation. This stance does not mean that the genetic counselor cannot mention spirituality or religion first. The key is to mention spirituality in a way that explores its *initial* inferential significance for the patient. As suggested above, mentioning spirituality in a list of other kinds of meaning may be the least intrusive strategy. If spirituality or religion is important to the counselee, then they will hopefully interpret this as an opportunity to become more directive in the conversation.

How would the responsibility model have handled Debbie's case in reference to the religious comments made? Coordination of meanings requires a navigation and negotiation between perspectives. A challenging first step for a genetic counselor – especially one uncomfortable talking about religion – is to continue navigating between perspectives after a religious claim has been made. To paraphrase Richard Rorty, religion can function as a conversation stopper but in genetic counseling it need not work this way.[62] In Chap. 3, I demonstrated in the concept of anaphora how navigation is possible with minimal understanding of what the other person means. Thus follow-up questions might entail "What's significant about *that* (God's will in all outcomes) for you?" The point here is that a genetic counselor does not have to adopt the idiom of the patient to navigate with the patient. Another important function of navigation is to track what Debbie has done in substituting talk of 'miraculous gift' for talk of babies and fetuses. "You have 1/106 risk of giving birth to *a miraculous gift* with Down syndrome" is a very important change in *P*. The genetic counselor should try to find out what the inferential significance of this move is. Brandom's explanation of how substitutions contribute to meaning is important in this situation.

The more difficult task for the responsibility model is to *negotiate* perspectives with Debbie. For example, would it be acceptable to press Debbie on her statement with this follow-up: "Are you saying that any decision you make will be the right one"? From the responsibility perspective, this measure would be acceptable because it allows Debbie to hear how someone else interprets her statement. This negotiation between perspectives is happening as part of the process of facilitating a responsible decision. The two goals of the responsibility account are both being pursued at his juncture in Debbie's case. If Debbie needs privacy at this moment, then she is expected to ask for it. What needs to be underscored in the negotiation process is that the genetic counselor must proceed with great caution because of his or her authority status. Mary White articulates this well in suggesting that all challenges to the patient need to be done with an affirming attitude similar to those of trusted friends, teachers and therapists.

---

[62] Richard Rorty, *Philosophy and Social Hope* (London, England; New York, N.Y., USA: Penguin, 1999).

The teaching model's objections are not rehearsed here because they are the same as those levied against the psychotherapeutic model. The psychotherapeutic model would endorse the responsibility model navigation of perspectives but would question its negotiation process. Because the counselee's perspective needs bolstering, the negotiation process threatens to undermine the autonomy of the patient. One of the underlying assumptions of the psychotherapeutic model is that the counselee is in a weaker position relative to the genetic counselor. This position is justifiable, but there are several ways to respond to this. Whereas Kessler and Weil seek to strengthen the whole person through empathy and unconditional positive regard, the responsibility model seeks to strengthen understanding and decision making through dialogical process. Its pragmatic emphasis stands out against the psychotherapeutic model's emphasis on changing psychological states.

The responsibility model provides the most adequate resources for responding to Debbie's case. To use Peters' image, it is the pragmatic middle ground in understanding what communication can achieve in this circumstance. On the one hand, the teaching model exhibits the frustration of having inadequate discursive resources to truly understand one another. In response to this limitation, it underestimates the amount of understanding that can be achieved between the genetic counselor and the patient. As a result, the HCP seeks primarily to transfer objective information in as clear and comprehensive manner as possible. Other communicative resources such as religious vocabulary are highly subjective and susceptible to misinterpretation. The genetic counselor must respect the gulf between perspectives and rely only on what can be sent and received effectively. On the other hand, the psychotherapeutic model overestimates the amount of understanding that can be achieved between patient and HCP. Its primary goal is to understand the other person rather than understand the conceptual contents with the other person. The shortcomings of relying on the empathic stance become perspicuous when addressing spirituality. Kessler's opening words to the chapter have a ring of inauthenticity that represent an unhelpful doubling of a perspective. Adopting the stance of the patient in this way begs for abuse and fails to see the value in letting the differences inform understanding and decision making. The strength of the responsibility model is that it acknowledges the value of the different perspective and focuses on how a shared responsibility to dialogical processes can coordinate meanings across perspectives. Unlike the teaching model, it does not restrict the meanings of genetic information. Religious meanings are fair game. Instead, this model commits to navigating and negotiating the different perspectives to support responsible decision making.

Summary

This chapter has tried to accomplish two goals. Most of the attention has been focused on assessing the prospect of instating a formal spiritual assessment in genetic counseling. In this process, I offered a new definition of spirituality that includes religion and medicine within it. Next, I surveyed reasons for undertaking spiritual assessment and rehearsed challenges to these reasons. The third subsection involved a detailed analysis of two studies that addressed the prospect of spiritual assessment in genetic counseling. Finally, after evaluating the benefits and harms of standardized spiritual assessment, I concluded that the potential harms outweighed the potential benefits.

The second goal of this chapter was to examine how the three genetic counseling models would address spirituality in Debbie's case. Each model's position was explained and then challenged by the other two models. In the end, the responsibility model was considered to be the most adequate of three models in addressing spiritual and religious concerns.

ERRATUM

Normative and Pragmatic Dimensions of Genetic Counseling

Negotiating Genetics and Ethics

Joseph B. Fanning

© Springer International Publishing Switzerland 2016
J.B. Fanning, *Normative and Pragmatic Dimensions of Genetic Counseling*,
Philosophy and Medicine 121, DOI 10.1007/978-3-319-44929-6

DOI 10.1007/978-3-319-44929-6_6

The original version of this book was inadvertently published with incorrect volume number. The correct volume number is 124. This has been updated in this volume.

The updated original online version for this book can be found at
http://dx.doi.org/10.1007/978-3-319-44929-6

© Springer International Publishing Switzerland 2016
J.B. Fanning, *Normative and Pragmatic Dimensions of Genetic Counseling*,
Philosophy and Medicine 124, DOI 10.1007/978-3-319-44929-6_6

Conclusion

This study brought together the very practical concern of talking with patients about genetic information with a highly theoretical discussion about communication and meaning. Drawing from the literature on genetic counseling, three models were introduced and then elaborated in several ways. An attempt was made to locate the models within larger streams of philosophical and theological thought. The teaching and psychotherapeutic models of genetic counseling were located within the what Peters calls the spiritualist tradition and were implicated in the technical and therapeutic visions of communication respectively. The responsibility model was elaborated in terms of an alternative vision of spirit rooted in Hegelian insights and underwritten by Robert Brandom's pragmatic theory of communication. By applying the models to the concerns around nondirectiveness and spiritual assessment, I tried to demonstrate how the responsibility model provides a better set of expressive resources about communication to guide practitioners.

Throughout the project Debbie's case served as a site for testing the adequacy of the models and as a source for examples. Chapters. 2 and 3 referred intermittently to her case as the models were being developed; Chaps. 4 and 5 invested significant attention to the case as a way to evaluate the adequacy of the models. Much more could have been said. For example, the issue around communicating probabilities has been the focus of significant research. An unfortunate consequence of focusing on a prenatal case is that it reinforced the perception that genetic counseling is exclusively about reproductive concerns. A few other examples were used to acknowledge the rapid expansion of genetics into almost all areas of medicine.

In Chap. 2, I introduced the teaching and psychotherapeutic models of genetic counseling. The assumption of this chapter was that these models implicitly relied on accounts of general communication to support their theses. John Durham Peters' story about communication in the West highlighted two main traditions: the spiritualist tradition and the embodiment tradition. He identified two visions of communication in the U.S. that have inherited the problematics of the spiritualist tradition. The technical and therapeutic visions of communication have, on Peters' view, dominated the U.S. since World War II, the same period when formative attitudes about genetic counseling were being pioneered. I proposed that the teaching model of genetic counseling incorporated many of the views of the technical vision; and that the psychotherapeutic model of genetic counseling was heir to the therapeutic vision of communication. Locating the models within broader views of communication allowed more explicit theses to be identified and ultimately evaluated in reference to the specific tenets of each model.

Chapter 3 consisted of a large-scale constructive move. I began by acknowledging Mary White's work that argues for placing sociality and responsibility at the center of genetic counseling, and that proposes the key concept of dialogical counseling. She developed these terms with the help of H. Richard Niebuhr's work in *The Responsible Self.* I elaborated and extended her stance in what I called responsibility model. As an heir to what I termed the embodiment tradition, the counterpart to the spiritualist tradition in Peters' account, the responsibility model embraces communication as a social practice that allows embodied selves to coordinate meanings across different perspectives. Hegel's insights into reciprocal recognition, sociality and historicity were acknowledged as the formative ideas in this tradition. To provide a more detailed footing for the responsibility model, I introduced and explicated Robert Brandom's pragmatic theory of communication. Using Brandom's theory to underwrite the responsibility model was in some respects an attempt to get back to the basics. His deontic scorekeeping account describes what happens in rudimentary conversations and what is required in normative practices to grasp a conceptual content. At the center of his model is the concept of dialogue. His detailed work provides White's dialogical counseling with sophisticated expressive resources that allowed the theses of the responsibility model to be elaborated.

The third chapter represented a shift from developing theoretical models to applying them. I examined one of the central values in genetic counseling, nondirectiveness. I demonstrated that nondirectiveness is a contested value with an important history. Nondirectiveness could best be understood as a corrective to the eugenic policies of the U.S. and Europe in the first of half of the twentieth century. Against this historical backdrop, I examined how each of the models specified and applied nondirectiveness. The teaching model defined it as the patient's right to noninterference with decision making. This view was justified with inadequate notions of pedagogical neutrality and autonomy. The psychotherapeutic model defined nondirectiveness as the promotion of autonomy and suggested that counseling skills should be actively used to encourage the patient. This view failed to maintain a meaningful distinction between directiveness and nondirectiveness by defining the former too narrowly and the latter too broadly. Resisting calls to jettison nondirectiveness as a defining value, I showed how the responsibility model provides a more nuanced understanding of the directive and nondirective stances that occur within genetic counseling sessions.

The final chapter addressed the prospects of spiritual assessment in genetic counseling. The interest by HCPs in spirituality has grown in the past fifteen years and the interest has spread to genetic counseling. I analyzed a representative definition of spirituality and concluded that a new definition was needed. A new definition was presented that used Daniel Sheridan's distinctions between culture, spirituality, and religion. His taxonomy provided a way to locate spirituality as mode of culture and to further divide spirituality into subspecies that included religion. This definition provided theoretical backing for the medical literature's preference for defining spirituality more broadly than religion. The next section presented reasons for and against spiritual assessment as a general proposal for HCPs. Richard Sloan's arguments against the partnership of medicine and religion provided needed push back for researchers who too quickly assume that spiritual assessment will improve patient care. In the genetic counseling context, very few studies have been undertaken in this area. Two studies that explored the possibility of spiritual assessment in genetic counseling were analyzed and evaluated. I concluded from these studies and the preceding arguments that a standardized spiritual assessment had more potential for harm than benefit. The final section of this chapter returned to Debbie's case to evaluate the adequacy of the three model's response to the religious concerns. I concluded that the responsibility model provided the most adequate model for addressing Debbie's religious concerns.

Implications

Genetic Counseling and Professional Communication

An obvious hope of this project is to have an effect on the practice of genetic counseling. This purpose is not based on the assessment that most genetic counselors are performing poorly. To the contrary, my limited contact with genetic counseling gave me the impression that they do a difficult job well. Nor do I pretend to have the know-how required to navigate and negotiate in the patient education room. Thus, an ambitious theoretical project like this one hopes to serve the more modest practical aim of supplementing knowing how with a knowing that. In other words, I tried to make explicit what I think many HCPs already do when they undertake genetic counseling. Nonetheless, some models are better than others, and the better ones might help in the process of training better counselors. I have tried to show that the responsibility model is better than the two dominant alternatives.

What has been learned in the present inquiry can be easily extended to other forms of health care communications and profession/client communication more generally. Several insights have been discovered in this project. Brandom's deontic scorekeeping model demonstrates how dialogue is at the root of grasping a conceptual content. Against this backdrop, professional communications such as genetic counseling can be seen as late developments in linguistic practices that have structural challenges. It appears as though the professional does not need to understand the esoteric information from the client's point of view. Brandom's model reminds us that the professional needs the client's understanding to grasp conceptual content in a particular context. One of the great challenges professionals have when talking to clients is to engage each person as a new dialogue partner for coordinating meanings. Another insight from the responsibility model is the distinction between navigating and negotiating perspectives. Many service professionals probably have an awareness of how they navigate a conversation but less awareness about how they negotiate perspectives within a conversation. Brandom's theory gives new resources for discourse analysis in this area. Third, the myth of professional neutrality is dismissed in this project. The myth of neutrality is the position that providing objective information is a neutral act. If Brandom's argument is accepted that linguistic practice is a fully normative practice, then the claim of neutrality is not credible. This insight expands current notions of professional and shared responsibility. Finally, the notion of shared decision making is an established domain in medical ethics but has received little attention in terms of communication theory. The weight Brandom places on dialogical processes makes his theory compatible with these ethical pursuits.

Medicine and Spirituality

The last chapter in this project has great relevance to a growing debate in medicine and the broader culture about the role spirituality should play in medicine. The definition of spirituality offered in this project has the potential to change the way we think about spheres of culture such as religion and medicine. Sheridan's broad notion of spirituality allows us to see both religion and medicine "as modes of culture in which human beings transform the problematic of the human predicament"; at the same time, the taxonomy allows for distinctions to be made between spiritualities. In an age when religion and science are simplistically pitted against one another, it is important to have categories that allow us to see their similarities and differences. Sheridan's framework can acknowledge that HCPs and patients both live in a world mediated by several modes of culture that shape their actual attitudes within clinical situations. These modes bring a variety of meanings into the health care setting that have to be coordinated across perspectives. If I think genes are the *Language of God*,[1] then an utterance of *P* means something different in my mouth than it does in your ears. Learning to talk about genetics will sometimes mean learning to talk about religion.

---

[1] Francis S. Collins, *The Language of God : A Scientist Presents Evidence for Belief* (New York: Free Press, 2006).

Appendix

Table 1 Example of question for the hope approach to spiritual assessment

H: Sources of **h**ope, meaning, comfort, strength, peace, love and connection
We have been discussing your support systems. I was wondering, what is there in your life that gives you internal support?
What are your sources of hope, strength, comfort and peace?
What do you hold on to during difficult times?
What sustains you and keeps you going?
For some people, their religious or spiritual beliefs act as a source of comfort and strength in dealing with life's ups and downs; is this true for you?
If the answer is "Yes," go on to O and P questions.
If the answer is "No," consider asking: Was it ever? If the answer is "Yes," ask: What changed?
O: **O**rganized religion
Do you consider yourself part of an organized religion?
How important is this to you?
What aspects of your religion are helpful and not so helpful to you?
Are you part of a religious or spiritual community? Does it help you? How?
P: **P**ersonal spirituality/**p**ractices
Do you have personal spiritual beliefs that are independent of organized religion? What are they?
Do you believe in God? What kind of relationship do you have with God?
What aspects of your spirituality or spiritual practices do you find most helpful to you personally? (e.g., prayer, meditation, reading scripture, attending religious services, listening to music, hiking, communing with nature)
E: **E**ffects on medical care and end-of-life issues
Has being sick (or your current situation) affected your ability to do the things that usually help you spiritually? (Or affected your relationship with God?)
As a doctor, is there anything that I can do to help you access the resources that usually help you?
Are you worried about any conflicts between your beliefs and your medical situation/care/decisions?

(continued)

Table 1 (continued)

Would it be helpful for you to speak to a clinical chaplain/community spiritual leader?
Are there any specific practices or restrictions I should know about in providing your medical care? (e.g., dietary restrictions, use of blood products)
If the patient is dying: How do your beliefs affect the kind of medical care you would like me to provide over the next few days/weeks/month

Bibliography

Anandarajah, G., and E. Hight. 2001. Spirituality and medical practice: Using the hope questions as a practical tool for spiritual assessment. *American Family Physician* 63(1): 81–9.

Aquinas, Thomas, and Dominicans. 1981. English Province. *Summa Theologica*. Complete English ed. Westminster: Christian Classics.

Armstrong, D., S. Michie, and T. Marteau. 1998. Revealed identity: A study of the process of genetic counselling. *Social Science and Medicine* 47(11): 1653–8.

Asch, A. 1998. Distracted by disability. The "difference" of disability in the medical setting. *Cambridge Quarterly of Healthcare Ethics* 7(1): 77–87.

Asch, A. 2003. Disability equality and prenatal testing: Contradictory or compatible? *Florida State Universtiy Law Review* 30(2): 315–42.

Baker, Diane L., Jane L. Schuette, Wendy R. Uhlmann, and NetLibrary Inc. 1998. *A guide to genetic counseling*. New York: Wiley-Liss.

Baumiller, R.C. 1974. Ethical issues in genetics. *Birth Defects Original Article Series* 10(10): 297–300.

Benkendorf, J.L., M.B. Prince, M.A. Rose, A. De Fina, and H.E. Hamilton. 2001. Does indirect speech promote nondirective genetic counseling? Results of a sociolinguistic investigation. *American Journal of Medical Genetics* 106(3): 199–207.

Berkowitz, R.L., J. Roberts, and H. Minkoff. 2006. Challenging the strategy of maternal age-based prenatal genetic counseling. *The Journal of the American Medical Association* 295(12): 1446–8.

Bernhardt, B.A., B.B. Biesecker, and C.L. Mastromarino. 2000. Goals, benefits, and outcomes of genetic counseling: Client and genetic counselor assessment. *American Journal of Medical Genetics* 94(3): 189–97.

Bosk, Charles. 1993. The workplace ideology of genetic counselors. In *Prescribing our future: Ethical challenges in genetic counseling*, ed. D. M. Bartels, B. LeRoy and Arthur L. Caplan. New York: Aldine de Gruyter, xii, 186 p.

Botkin, J.R. 1995. Federal privacy and confidentiality. *The Hastings Center Report* 25(5): 32–38.

Boyle, P. J. 2004. Genetics and pastoral counseling: A special report. *Second Opinion (Chicago)* (11): 4–56.

Brandom, Robert. 1994. *Making it explicit: Reasoning, representing, and discursive commitment*. Cambridge, MA: Harvard University Press.

Brandom, Robert. 2000. Facts, norms, and normative facts: Reply to Habermas. *European Journal of Philosophy* 8(3): 356–74.

Brandom, Robert. 2002. *Tales of the mighty dead: Historical essays in the metaphysics of intentionality*. Cambridge, MA: Harvard University Press.

Brandom, Robert. 2004. Hermeneutic practice and theories of meaning. *SATS – Nordic Journal of Philosophy* 5(1): 5–26.

Bussing, A., T. Ostermann, and H.G. Koenig. 2007. Relevance of religion and spirituality in German patients with chronic diseases. *International Journal of Psychiatry in Medicine* 37(1): 39–57.

Churchill, L.R., and D. Schenck. 2005. One cheer for bioethics: Engaging the moral experiences of patients and practitioners beyond the big decisions. *Cambridge Quarterly of Healthcare Ethics* 14(4): 389–403.

Clarke, A., E. Parsons, and A. Williams. 1996. Outcomes and process in genetic counselling. *Clinical Genetics* 50(6): 462–9.

Clayton, E.W. 2006. The web of relations: Thinking about physicians and patients. *Yale Journal of Health Policy, Law, and Ethics* 6(2): 465–77; discussion 79–502.

College, Sarah Lawrence. 2008. *Human genetics 2007–2008 Courses* [website]. Bronxville: Sarah Lawrence College. Available from http://www.slc.edu/human-genetics/Courses.php. Accessed 11 Jan 2008.

Collins, Francis S. 2006. *The language of God: A scientist presents evidence for belief*. New York: Free Press.

Counseling, Ad Hoc Committee on Genetic. 1975. Genetic counseling. *American Journal of Human Genetics* 27(2): 240–242.

Davidson, Donald. 1963. Actions, reasons, and causes. *The Journal of Philosophy* 60(23): 685–700.

de Crespigny, L. 2003. Words matter: Nomenclature and communication in perinatal medicine. *Clinics in Perinatology* 30(1): 17–25.

Dice, Lee. 1952. Genetic counseling. *American Journal of Human Genetics* 4(4): 332–46.

Douglas, Mary. 1992. *Risk and blame: Essays in cultural theory*. London/New York: Routledge.

Dyson, J., M. Cobb, and D. Forman. 1997. The meaning of spirituality: A literature review. *Journal of Advanced Nursing* 26(6): 1183–8.

Ehman, J.W., B.B. Ott, T.H. Short, R.C. Ciampa, and J. Hansen-Flaschen. 1999. Do patients want physicians to inquire about their spiritual or religious beliefs if they become gravely ill? *Archives of Internal Medicine* 159(15): 1803–6.

Eisinger, F. 2007. Prophylactic mastectomy: Ethical issues. *British Medical Bulletin* 81–82: 7–19.

Ellington, L., D. Roter, W.N. Dudley, B.J. Baty, R. Upchurch, S. Larson, J.E. Wylie, K.R. Smith, and J.R. Botkin. 2005. Communication analysis of Brca1 genetic counseling. *Journal of Genetic Counseling* 14(5): 377–86.

Ellington, L., B.J. Baty, J. McDonald, V. Venne, A. Musters, D. Roter, W. Dudley, and R.T. Croyle. 2006. Exploring genetic counseling communication patterns: The role of teaching and counseling approaches. *Journal of Genetic Counseling* 15(3): 179–89.

Ellington, L., K.M. Kelly, M. Reblin, S. Latimer, and D. Roter. 2011. Communication in genetic counseling: Cognitive and emotional processing. *Health Communication* 26(7): 667–675.

Farrelly, E., M.K. Cho, L. Erby, D. Roter, A. Stenzel, and K. Ormond. 2012. Genetic counseling for prenatal testing: Where is the discussion about disability? *Journal of Genetic Counseling* 21(6): 814–824.

Fine, Beth. 1993. The evolution of nondirectiveness in genetic counseling and implications of the human genome project. In *Prescribing our future: Ethical challenges in genetic counseling*, ed. D.M. Bartels, Bonnie LeRoy, and Arthur L. Caplan. New York: Aldine de Gruyter.

Freire, Paulo. 1970. *Pedagogy of the oppressed*. New York: Herder and Herder.

Geertz, Clifford. 1973. *The interpretation of cultures; selected essays*. New York: Basic Books.

Giroux, Henry A. 1997. *Pedagogy and the politics of hope: Theory, culture, and schooling: A critical reader*, The edge, critical studies in educational theory. Boulder: Westview Press.

Habermas, Jürgen. 2000. From Kant to Hegel: On Robert Brandom's pragmatic philosophy of language. *European Journal of Philosophy* 8(3): 322–55.

Habermas, Jürgen, Ciaran Cronin, and Max Pensky. 2006. *Time of transitions*. Cambridge/Malden: Polity.

Harmon, Amy. 2007, May 9. Prenatal test puts down syndrome in hard focus. *New York Times* 1.
Harrison, M.O., C.L. Edwards, H.G. Koenig, H.B. Bosworth, L. Decastro, and M. Wood. 2005. Religiosity/spirituality and pain in patients with sickle cell disease. *The Journal of Nervous and Mental Disease* 193(4): 250–7.
Heffner, L.J. 2004. Advanced maternal age – How old is too old? *The New England Journal of Medicine* 351(19): 1927–9.
Hegel, Georg Wilhelm Friedrich, and Peter Crafts Hodgson. 1988. *Lectures on the philosophy of religion: The lectures of 1827*, one-volume edition. Berkeley: University of California Press.
Hegel, Georg Wilhelm Friedrich, Arnold Vincent Miller, and J.N. Findlay. 1977. *Phenomenology of spirit*. Oxford: Clarendon Press.
Hook, E.B., P.K. Cross, and D.M. Schreinemachers. 1983. Chromosomal abnormality rates at amniocentesis and in live-born infants. *The Journal of the American Medical Association* 249(15): 2034–8.
Hsia, Y. Edward. 1979. The genetic counselor as information giver. In *Genetic counseling: Facts, values, and norms*, vol. 15, ed. Alexander Morgan Capron, 169–86. New York: Alan R. Liss.
Kaldjian, L.C., J.F. Jekel, and G. Friedland. 1998. End-of-life decisions in HIV-positive patients: The role of spiritual beliefs. *AIDS* 12(1): 103–7.
Kant, Immanuel. 1991. An answer to the question: What is enlightenment? In *Kant: Political writings*, vol. xv, ed. Hans Siegbert Reiss. Cambridge/New York: Cambridge University Press, 311 p.
Kay, Lily E. 2000. *Who wrote the book of life? A history of the genetic code*, Writing science. Stanford: Stanford University Press.
Kelly, Patricia T. 1977. *Dealing with dilemma: A manual for genetic counselors*, Heidelberg science library. New York: Springer-Verlag.
Kenen, R., and J. Peters. 2001. The colored, eco-genetic relationship map (cegrm): A conceptual approach and tool for genetic counseling research. *Journal of Genetic Counseling* 10(4): 289–309.
Kessler, Seymour. 1979. *Genetic counseling: Psychological dimensions*. New York: Academic Press.
Kessler, S. 1981. Psychological aspects of genetic counseling: Analysis of a transcript. *American Journal of Medical Genetics* 8(2): 137–53.
Kessler, S. 1997a. Psychological aspects of genetic counseling. Ix. Teaching and counseling. *Journal of Genetic Counseling* 6(3): 287–95.
Kessler, S. 1997b. Psychological aspects of genetic counseling. Xi. Nondirectiveness revisited. *American Journal of Medical Genetics* 72(2): 164–71.
Kessler, S. 1998. Psychological aspects of genetic counseling: Xii. More on counseling skills. *Journal of Genetic Counseling* 7(3): 263–78.
Kessler, S. 2001. Psychological aspects of genetic counseling. Xiv. Nondirectiveness and counseling skills. *Genetic Testing* 5(3): 187–91.
Kessler, S., and E.K. Levine. 1987. Psychological aspects of genetic counseling. Iv. The subjective assessment of probability. *American Journal of Medical Genetics* 28(2): 361–70.
Kevles, Daniel J., and Leroy E. Hood. 1992. *The code of codes: Scientific and social issues in the human genome project*. Cambridge, MA: Harvard University Press.
King, D.E., and B. Bushwick. 1994. Beliefs and attitudes of hospital inpatients about faith healing and prayer. *Journal of Family Practice* 39(4): 349–52.
Koenig, H.G. 2001. Religion and medicine Iv: Religion, physical health, and clinical implications. *International Journal of Psychiatry in Medicine* 31(3): 321–36.
Korenberg, J.R., X.N. Chen, R. Schipper, Z. Sun, R. Gonsky, S. Gerwehr, N. Carpenter, C. Daumer, P. Dignan, C. Disteche, et al. 1994. Down syndrome phenotypes: The consequences of chromosomal imbalance. *Proceedings of the National Academy of Sciences of the United States of America* 91(11): 4997–5001.
Lewis, L.J. 2002. Models of genetic counseling and their effects on multicultural genetic counseling. *Journal of Genetic Counseling* 11(3): 193–212.

Lippman-Hand, A., and F.C. Fraser. 1979a. Genetic counseling--the postcounseling period: I. Parents' perceptions of uncertainty. *American Journal of Medical Genetics* 4(1): 51–71.

Lippman-Hand, A., and F.C. Fraser. 1979b. Genetic counseling: Parents' responses to uncertainty. *Birth Defects Original Article Series* 15(5C): 325–39.

Lippman-Hand, A., and F.C. Fraser. 1979c. Genetic counseling: Provision and reception of information. *American Journal of Medical Genetics* 3(2): 113–27.

Locke, John, and P.H. Nidditch. 1979. *An essay concerning human understanding*. The Clarendon Edition of the works of John Locke. Oxford/New York: Clarendon Press/Oxford University Press.

Maaskant, M.A., M. van den Akker, A.G. Kessels, M.J. Haveman, H.M. van Schrojenstein Lantman-de Valk, and H.F. Urlings. 1996. Care dependence and activities of daily living in relation to ageing: Results of a longitudinal study. *Journal of Intellectual Disability Research* 40(Pt 6): 535–43.

Marks, Joan. 1993. The training of genetic counselors: Origins of a psychosocial model. In *Prescribing our future: Ethical challenges in genetic counseling*, vol. xii, ed. D.M. Bartels, B. LeRoy, and Arthur L. Caplan. New York: Aldine de Gruyter, 186p.

Maugans, T.A., and W.C. Wadland. 1991. Religion and family medicine: A survey of physicians and patients. *Journal of Family Practice* 32(2): 210–3.

Mazzoni, D.S., R.S. Ackley, and D.J. Nash. 1994. Abnormal pinna type and hearing loss correlations in down's syndrome. *Journal of Intellectual Disability Research* 38(Pt 6): 549–60.

McSherry, W., and K. Cash. 2004. The language of spirituality: An emerging taxonomy. *International Journal of Nursing Studies* 41(2): 151–61.

Michie, S., F. Bron, M. Bobrow, and T.M. Marteau. 1997. Nondirectiveness in genetic counseling: An empirical study. *American Journal of Human Genetics* 60(1): 40–7.

Moreira-Almeida, A., and H.G. Koenig. 2006. Retaining the meaning of the words religiousness and spirituality: A commentary on the Whoqol Srpb Group's "a cross-cultural study of spirituality, religion, and personal beliefs as components of quality of life" (62: 6, 2005, 1486–1497). *Social Science and Medicine* 63(4): 843–5.

Motulsky, Arno. 2004. 2003 Ashg Award for excellence in human genetics education: Introductory remarks for Joan Marks. *American Journal of Human Genetics* 74: 393–94.

Nicolaides, K.H., F.A. Chervenak, L.B. McCullough, K. Avgidou, and A. Papageorghiou. 2005. Evidence-based obstetric ethics and informed decision-making by pregnant women about invasive diagnosis after first-trimester assessment of risk for trisomy 21. *American Journal of Obstetrics and Gynecology* 193(2): 322–6.

Niebuhr, H. Richard. 1963. *The responsible self; an essay in Christian moral philosophy*, 1st ed. New York: Harper & Row.

Nussbaum, Robert L., Roderick R. McInnes, Huntington F. Willard, Margaret W. Thompson, and James S. Thompson. 2001. *Thompson & Thompson genetics in medicine*, 6th ed. Philadelphia: Saunders.

O'Dea, Thomas F. 1966. *The sociology of religion*, Foundations of modern sociology series. Englewood Cliffs: Prentice-Hall.

Oyama, O., and H.G. Koenig. 1998. Religious beliefs and practices in family medicine. *Archives of Family Medicine* 7(5): 431–5.

Parens, Erik, and Adrienne Asch. 2000. *Prenatal testing and disability rights*, Hastings center studies in ethics. Washington, D.C.: Georgetown University Press.

Parens, E., and A. Asch. 2003. Disability rights critique of prenatal genetic testing: Reflections and recommendations. *Mental Retardation and Developmental Disabilities Research Reviews* 9(1): 40–7.

Paul, Diane B. 1995. *Controlling human heredity, 1865 to the present*, The control of nature. Atlantic Highlands: Humanities Press.

Peregrin, Jaroslav. 2001. *Meaning and structure: Structuralism of (post)analytic philosophers*, Ashgate new critical thinking in philosophy. Aldershot/Burlington: Ashgate.

Peters, Ted. 1996. *For the love of children: Genetic technology and the future of the family*, Family, religion, and culture, 1st ed. Louisville: Westminster John Knox Press.

Peters, John Durham. 1999. *Speaking into the air: A history of the idea of communication*. Chicago: University of Chicago Press.

Peters, J.A., R. Kenen, R. Giusti, J. Loud, N. Weissman, and M.H. Greene. 2004. Exploratory study of the feasibility and utility of the colored eco-genetic relationship map (Cegrm) in women at high genetic risk of developing breast cancer. *American Journal of Medical Genetics. Part A* 130(3): 258–64.

Peters, J.A., L. Hoskins, S. Prindiville, R. Kenen, and M.H. Greene. 2006. Evolution of the colored Eco-genetic relationship map (Cegrm) for assessing social functioning in women in hereditary breast-ovarian (Hboc) families. *Journal of Genetic Counseling* 15(6): 477–89.

Radford, Gary P. 2005. *On the philosophy of communication*, Wadsworth philosophical topics. Belmont: Thomson Wadsworth.

Rapp, R. 1988. Chromosomes and communication: The discourse of genetic counseling. *Medical Anthropology Quarterly* 2(2): 143–57.

Rebbeck, T.R., T. Friebel, H.T. Lynch, S.L. Neuhausen, L. van't Veer, J.E. Garber, G.R. Evans, S.A. Narod, C. Isaacs, E. Matloff, M.B. Daly, O.I. Olopade, and B.L. Weber. 2004. Bilateral prophylactic mastectomy reduces breast cancer risk in Brca1 and Brca2 mutation carriers: The prose study group. *Journal of Clinical Oncology* 22(6): 1055–62.

Reed, Sheldon Clark. 1955. *Counseling in medical genetics*. Philadelphia: Saunders.

Reis, L.M., R. Baumiller, W. Scrivener, G. Yager, and N.S. Warren. 2007. Spiritual assessment in genetic counseling. *Journal of Genetic Counseling* 16(1): 41–52.

Resta, R.G. 1997. Eugenics and nondirectiveness in genetic counseling. *Journal of Genetic Counseling* 6(2): 255–8.

Resta, R.G. 2005. Changing demographics of advanced maternal age (Ama) and the impact on the predicted incidence of down syndrome in the United States: Implications for prenatal screening and genetic counseling. *American Journal of Medical Genetics. Part A* 133(1): 31–6.

Resta, R.G., and S. Kessler. 2004. Commentary on Robin's a smile, and the need for counseling skills in the clinic. *American Journal of Medical Genetics. Part A* 126(4): 437–8; author reply 39.

Resta, R., B.B. Biesecker, R.L. Bennett, S. Blum, S.E. Hahn, M.N. Strecker, and J.L. Williams. 2006. A new definition of genetic counseling: National society of genetic counselors' task force report. *Journal of Genetic Counseling* 15(2): 77–83.

Rippentrop, E.A., E.M. Altmaier, J.J. Chen, E.M. Found, and V.J. Keffala. 2005. The relationship between religion/spirituality and physical health, mental health, and pain in a chronic pain population. *Pain* 116(3): 311–21.

Rogers, Carl R. 1951. *Client-centered therapy, its current practice, implications, and theory*. Boston: Houghton Mifflin.

Rogers, Carl R. 1961. *On becoming a person; a therapist's view of psychotherapy*. Boston: Houghton Mifflin.

Rorty, Richard. 1999. *Philosophy and social hope*. London/New York: Penguin.

Roter, D., L. Ellington, L.H. Erby, S. Larson, and W. Dudley. 2006. The genetic counseling video project (Gcvp): Models of practice. *American Journal of Medical Genetics. Part C, Seminars in Medical Genetics* 142(4): 209–20.

Rothenberg, Karen H., and Elizabeth J. Thomson. 1994. *Women and prenatal testing: Facing the challenges of genetic technology*, Women and health series. Columbus: Ohio State University Press.

Rothman, Barbara Katz. 1986. *The tentative pregnancy: Prenatal diagnosis and the future of motherhood*. New York: Viking.

Scharp, Kevin. 2003. Communication and content: Circumstances and consequences of the Habermas-Brandom debate. *International Journal of Philosophical Studies* 11(1): 43–61.

Schwartz, M.D., C. Hughes, J. Roth, D. Main, B.N. Peshkin, C. Isaacs, C. Kavanagh, and C. Lerman. 2000. Spiritual faith and genetic testing decisions among high-risk breast cancer probands. *Cancer Epidemiology, Biomarkers & Prevention* 9(4): 381–5.

Searle, John R. 1979. *Expression and meaning: Studies in the theory of speech acts*. Cambridge/New York: Cambridge University Press.

Sellars, Wilfrid. 1954. Some reflections on langauge games. *Philosophy of Science* 21(3): 204–28.

Shannon, Claude Elwood, and Warren Weaver. 1949. *The mathematical theory of communication*. Urbana: University of Illinois Press.

Sheridan, Daniel. 1986. Discerning difference: A taxonomy of culture, spirituality, and religion. *The Journal of Religion* 66(1): 37–45.

Sloan, Richard P. 2006. *Blind faith: The unholy alliance of religion and medicine*, 1st ed. New York: St. Martin's Press.

Sorenson, J.R. 1993. Genetic counseling: Values that have mattered. In *Prescribing our future: Ethical challengs in genetic counseling*, ed. D.M. bartels, 3–14. New York: Aldine De Gruyter.

Sorenson, J.R., C.M. Kavanagh, and M. Mucatel. 1981. Client learning of risk and diagnosis in genetic counseling. *Birth Defects Original Article Series* 17(1): 215–28.

Stefanek, M., P.G. McDonald, and S.A. Hess. 2005. Religion, spirituality and cancer: Current status and methodological challenges. *Psychooncology* 14(6): 450–63.

Stern, A. 2012. *Telling genes: The story of genetic counseling in America*. Baltimore: Johns Hopkins University Press.

Stout, Jeffrey. 1988. *Ethics after babel: The languages of morals and their discontents*. Boston: Beacon Press.

Stout, Jeffrey. 2004. *Democracy and tradition*, New Forum Books. Princeton: Princeton University Press.

Suter, S.M. 1998. Value neutrality and nondirectiveness: Comments on "future directions in genetic counseling". *Kennedy Institute of Ethics Journal* 8(2): 161–3.

Urbano, R.C., and R.M. Hodapp. 2007. Divorce in families of children with down syndrome: A population-based study. *American Journal of Mental Retardation* 112(4): 261–74.

van den Berg, M., D.R.M. Timmermans, J.H. Kleinveld, J.T.M. van Eijk, D.L. Knol, G. van der Wal, and J.M.G. van Vugt. 2007. Are counsellors' attitudes influencing pregnant women's attitudes and decisions on prenatal screening? *Prenatal Diagnosis* 27(6): 518–524.

Vanderbilt Center for Integrative Health. 2007. Nashville: Vanderbilt University Medical Center. Available from http://www.vanderbilthealth.com/integrativehealth/. Accessed 17 Dec 2007.

Veach, P.M., D.M. Bartels, and B.S. LeRoy. 2007. Coming full circle: A reciprocal-engagement model of genetic counseling practice. *Journal of Genetic Counseling* 16(6): 713–728.

Wachholtz, A.B., M.J. Pearce, and H. Koenig. 2007. Exploring the relationship between spirituality, coping, and pain. *Journal of Behavioral Medicine* 30(4): 311–8.

Weil, Jon. 2000. *Psychosocial genetic counseling*, Oxford monographs on medical genetics; No. 41. New York: Oxford University Press.

Weil, J. 2003. Psychosocial genetic counseling in the post-nondirective era: A point of view. *Journal of Genetic Counseling* 12(3): 199–211.

Weil, J., K. Ormond, J. Peters, K. Peters, B.B. Biesecker, and B. LeRoy. 2006. The relationship of nondirectiveness to genetic counseling: Report of a workshop at the 2003 Nsgc annual education conference. *Journal of Genetic Counseling* 15(2): 85–93.

Wertz, D.C., and J.C. Fletcher. 1988. Attitudes of genetic counselors: A multinational survey. *American Journal of Human Genetics* 42(4): 592–600.

Wertz, D.C., J.R. Sorenson, and T.C. Heeren. 1986. Clients' interpretation of risks provided in genetic counseling. *American Journal of Human Genetics* 39(2): 253–64.

White, M.T. 1997. "Respect for autonomy" in genetic counseling: An analysis and a proposal. *Journal of Genetic Counseling* 6(3): 297–313.

White, M.T. 1999. Making responsible decisions. An interpretive ethic for genetic decision making. *The Hastings Center Report* 29(1): 14–21.

Index

A
Abandoning, 36, 129
Abortion, 6, 50, 79, 83, 97, 103
Acceptance, 9, 16, 17, 26, 33, 34, 84, 86, 92, 97, 109, 114, 131, 138
Acknowledge, 13, 16, 20, 24–26, 29–31, 37, 38, 43–45, 48, 49, 51, 56, 59, 60, 70, 71, 76, 80, 82, 86, 89–91, 93, 98, 99, 101–103, 105, 106, 114, 115, 119, 122, 128, 132, 135, 138
Action, 1, 12, 17, 20, 43, 48–50, 59, 68–70, 73, 86, 90, 91, 94, 98, 99, 101, 106–108, 112, 117
Adapt, 12, 37, 47, 61, 84, 86
Adaptation, 10, 12
Amniocentesis, 1, 5, 6, 30, 51, 61, 67, 69, 75, 76, 82, 86, 91, 94, 95, 97, 98, 102, 122, 126
Anaphora, 73, 74, 131
Anaphoric, 73, 74
Anxiety, 38, 44, 61, 68, 95, 108, 123, 127
Aquinas, 15, 17, 52
Aristotle, 54
Articulation, 3, 4, 12, 14, 24, 27, 28, 39, 41, 52, 55, 68, 80, 85, 86, 88, 89, 92, 99, 101, 104, 106, 129, 131
Asch, A., 7, 62, 95
Ascription, 31, 40, 60, 65, 66, 101, 119
Atomistic, 22, 26, 29, 48, 88, 127
Attention, 6, 7, 16, 38, 43, 53, 57, 61, 62, 71, 74, 91, 94, 98, 104, 113, 120, 122, 132, 135, 138
Attribute, 29–31, 33, 42, 50, 54, 56–61, 64, 66, 68, 71, 95, 108, 109, 112, 116
Augustine, 7, 15, 17, 52
Authority, 3, 10, 25, 31, 45, 51, 54, 56, 61, 70, 72, 77, 82, 85, 86, 88, 94, 95, 97–99, 101, 102, 112, 121, 122, 126, 127, 129–131
Authorize, 56, 58, 94
Autonomy, social dimensions of, 94

B
Bacon, F., 17, 18
Belief, 3, 18, 26, 27, 29, 31, 34, 35, 42, 43, 47, 50, 53, 58, 60, 64, 66, 69, 70, 77, 93, 95, 97, 105, 106, 108–112, 115–118, 120–123, 125–127, 129, 130, 139, 140
Beneficence, 111
Benefit, 8, 18, 28, 31, 40, 51, 54, 83, 87, 95, 104, 108, 109, 111, 112, 119, 120, 123, 124, 133, 137
Bodily, 14–16, 22, 36, 37, 54, 68, 83, 95, 100, 109
Bracket, 41, 88, 127, 129
Brandom, R., 3, 7, 13, 47, 50, 51, 54–69, 72, 73, 75–78, 93, 94, 100, 131, 135–138
Bricoleur, 125
Buddhism, 107

C
Caplan, A.L., 27
Challenge, 2, 4, 12, 13, 18, 22, 23, 26, 28–30, 37, 42, 48, 54, 59–62, 67, 69–73, 78, 80, 82, 83, 86, 87, 95, 97, 98, 102, 104, 108, 110–112, 115, 116, 126, 128, 130, 131, 133, 137
Churchill, L.R., 71, 109, 125

© Springer International Publishing Switzerland 2016
J.B. Fanning, *Normative and Pragmatic Dimensions of Genetic Counseling*,
Philosophy and Medicine 124, DOI 10.1007/978-3-319-44929-6

Claiming, 3, 7, 9, 12, 14, 20, 22, 24, 26, 28, 30, 31, 35, 36, 39, 41, 42, 45, 47, 48, 50–52, 58–61, 63, 64, 66–68, 70, 74, 76, 78, 82, 85, 87, 89, 94, 106, 107, 111–113, 120, 122, 123, 131, 138
Client-centered, 2, 4, 10–12, 24, 26, 28–31, 33, 34, 37–39, 41–43, 45, 47–49, 51, 61, 63, 65, 70–72, 78, 80, 82, 84, 86, 88–92, 95, 99–102, 114, 115, 119, 120, 122, 124, 127, 129, 130, 137
Coercion, 8, 48, 70, 91–93, 101, 112
Cognitive, 10, 29–31, 38, 42, 62, 73, 83, 105, 108
Commitment, 3, 10–12, 26, 28, 30, 32, 38, 40–43, 49–51, 55–64, 67, 68, 70, 71, 73–75, 88, 89, 92–95, 97–100, 121–123, 126–130
Communication, v, 1–4, 7–45, 47, 51–67, 72, 73, 77–79, 84, 88, 89, 91–93, 99, 100, 103, 119, 124, 127, 130, 132, 135–138
Communicative, 2, 14, 15, 24, 32, 39, 40, 53, 55, 72, 132
Community, 21, 42, 49, 50, 60, 64, 81, 105, 106, 117, 128, 139, 140
Congruence, 32, 33, 35, 89
Constraint, 17, 18, 41, 48, 49, 54, 60, 68, 88, 90, 92, 97, 108, 109, 115, 126
Context, 10, 13, 22, 39, 49, 50, 55, 58, 61, 63, 64, 66, 67, 69, 72, 73, 75, 77, 82, 84, 85, 90, 105, 107, 120, 122, 123, 137
Conversational scorekeeping, 1, 47, 58, 59, 93, 96, 98
Coordinate, 8, 18, 37, 43, 47, 49, 52, 54, 63, 67, 72, 77, 95, 122, 124, 132, 136
Coping, 12, 38, 86, 111–113, 128
Critical reflection, 4
Culture, 13, 14, 18, 37, 38, 41, 49, 53, 65, 97, 106–110, 122, 125, 136–138

D

Decision making, 2, 6–8, 10–12, 29, 31, 42, 47–50, 55, 66–72, 75–77, 79, 80, 83, 84, 86–90, 92–103, 110–112, 114, 122–124, 127–132, 136, 138
Deductive, 61
Default-challenge structure, 59, 60
Deference, 61, 64, 82, 130
Deliberation, 5, 6, 29, 48, 70, 71, 98, 125, 126, 130
Desire, 19, 26, 33, 39, 45, 68, 69, 75, 76
Dialogical process, 45, 48, 50, 64–67, 72, 73, 75, 77, 94, 127, 130, 132, 138
Dialogical relation, 64–68, 70, 72, 73, 75, 77, 93, 102, 124, 130

Dialogue, 48, 53, 55, 64, 66, 70, 72, 73, 92, 136, 137
Difference, 2–4, 8, 11, 12, 18, 30, 41, 43, 45, 51, 53, 55, 56, 63, 65, 67, 73, 74, 77, 90, 92, 94, 101, 109, 117, 121, 124, 129, 132, 138
Dilemma, 44, 76, 107, 116
Directive, 84, 86, 91, 93–95, 98, 100–102, 126, 131, 136
Directiveness, 93
Directiveness, doxastic, practical, 88, 95, 98
Disability, 62, 83, 95, 109
Discourse, 14, 16, 20, 32, 73, 81, 138
Discursive, 4, 15, 17, 18, 42, 45, 48, 50, 51, 56, 60, 61, 93, 132
Dissemination, 53
Distance, 15–18, 20, 44, 51, 86, 117
Doubling, 18, 130, 132
Down syndrome, 5, 6, 42, 49–51, 53, 55, 58, 60, 62, 74, 86, 95, 96, 126, 131
Doxastic, 64, 66, 73, 93–95, 97–102

E

Embodiment, v, 4, 7, 8, 15, 23, 37, 51–54, 67, 77, 78, 90, 108, 135, 136
Embodiment, tradition, v, 4, 51–54, 77, 135, 136
Emotion, 11, 22, 24, 26–28, 34, 37, 38, 43, 44, 49, 68, 72, 73, 80, 87, 89, 90, 92, 93, 97, 98, 100–102, 105, 106, 108, 109, 111, 123, 129
Empathic, 33–36, 38, 41, 42, 45, 47, 51, 54, 55, 68, 70, 78, 90, 92, 99, 119, 127–130, 132
Empathy, 36, 42, 44, 79, 90, 130, 132
End-of-life, 71, 114, 117, 139
Entitlement, 16, 55–64, 69, 71, 76, 85, 87, 88, 95, 101, 121, 126, 128, 129
Ethical, 1, 2, 6, 7, 12, 49, 56, 70, 76, 78, 104, 138
Experience, 5, 11, 15–17, 24, 27, 32–39, 42–44, 54, 61, 63, 67, 72, 73, 80, 82, 87, 91, 94, 103, 105, 106, 119, 128
Explicit, 13, 15, 21, 25, 27, 32, 35, 36, 38, 43, 47, 50, 51, 53–56, 58–60, 63, 65–70, 72, 73, 76, 78, 85–88, 92, 94, 96, 100, 102, 116, 124, 129, 136, 137
Expressive resources, 4, 8, 13, 16, 45, 47, 54, 135, 136
Exterior, 14, 15, 52

F

Faith, 1, 44, 103, 104, 106, 110, 111, 119, 124
Fear, 33, 36, 68, 106, 116

Index

Feelings, 2, 22, 23, 27, 33, 35, 37–39, 42, 73, 82, 92, 100, 105, 106, 119
Fetus, 1, 49, 51, 54, 58, 63, 65, 66, 71, 76, 79, 95–97, 131
Framework, 4, 19, 43, 54, 73, 89, 92, 98, 107, 117, 125, 126, 128, 138

G

Gadamer, Hans-George. *See* Gadamerian
Gadamerian, 22, 63
Geertz, C., 107
Genealogy, 14, 18, 20
Genetic counseling, 1–45, 47–133, 135–138
Genetic counselor, 1, 2, 5, 6, 8, 9, 11, 20, 23–31, 34, 37–44, 47–49, 53–55, 58–65, 67–72, 74, 76, 77, 79, 80, 82, 84–88, 90, 91, 93–104, 111, 113, 114, 116–122, 124–132, 137
Genetics, 1–14, 18–32, 34–45, 47–51, 53–55, 58–106, 110, 111, 113–133, 135–138
Guilt, 29, 42, 73, 112

H

Habermas, J., 55, 78
Harm, 8, 18, 30, 42, 49, 53, 70, 83, 87, 97, 104, 111, 112, 120–122, 124, 133, 137
Healing, 109, 119
Hegel, G.W.F., 50, 52, 136
Hermeneutic, 3, 65, 66
History, intellectual; of genetic counseling, 3, 14, 27
Hope, 8, 18, 42, 43, 72, 105, 106, 116, 119, 128, 131, 137, 139
HOPE approach, 114, 118–121, 127, 128, 130
Hsia, Y.E., 3, 24–31, 71, 85–88, 129

I

Identity, 3, 4, 6, 7, 11, 14, 18, 20, 33, 35, 41, 42, 45, 50, 51, 60, 64, 66, 69, 70, 74, 76, 78, 84, 90, 91, 96, 99, 100, 102, 108, 110, 111, 114, 116, 117, 119–121, 125, 128–130, 135, 136
Ideology, 23, 90, 108
Implicit, 13, 31, 39, 53–55, 58, 60, 61, 65, 66, 68, 70–72, 76, 77, 85, 87, 88, 92, 96, 99, 102, 106, 112, 126, 129, 135
Incommensurable, 51, 65, 126
Incompatible commitments, 15, 31, 97, 116
Individual, 10, 12, 15, 16, 22, 35–37, 41–43, 47–49, 52, 53, 63–65, 69, 70, 80, 85, 88, 94, 97, 105, 108, 111, 118, 119, 122, 124
Inequality, 77
Inference, logical, material, 29, 59, 100
Inferential, 56, 59–61, 64, 66, 68, 71, 84, 88, 93–96, 100–102, 127, 131
Inheritance, intrapersonal, interpersonal, 7, 60
Intentional stance, 108–110, 125
Interest–patient's, best, 111, 112, 116, 118
Interiority, 14–19, 27, 29, 32–34, 36, 45, 52, 53, 68, 77, 99
Interpretation, 5, 7, 12, 14, 19–22, 26, 27, 29, 30, 36, 40, 43, 45, 48, 50, 51, 63–67, 69, 73–75, 85, 99, 103, 107, 108, 110, 117, 120, 123, 124, 126, 131

J

Judgment, 33, 43, 50, 51, 72, 85, 86
Justify, 13, 28, 41, 51, 55, 57, 59, 60, 71, 81, 84, 88, 96–98, 100, 104, 108, 111, 112, 119, 132, 136

K

Kant, I., 50, 76, 78
Kessler, S., 2, 3, 6, 7, 9–13, 23, 24, 27–29, 31, 36–41, 43–45, 49, 68, 72, 73, 75, 80, 84, 90–94, 101, 103, 128–130, 132
Kinematics, 56, 100

L

Levinas, E., 78
Linguistic, 3, 8, 13, 19–23, 30, 36, 50, 52, 55, 56, 58–61, 64, 66, 74, 78, 137, 138
Listening, 1, 26, 33, 44, 92, 139
Locke, J., 7, 15–17, 52
Logic, 13, 15, 29, 59, 76, 92, 100
Love, 103, 105, 106, 116, 128, 139

M

Manipulation, 31, 101, 112, 120, 122, 126
Mastectomy, 83, 90
Material inference, 59, 66, 75, 76
Maternal, 1, 6, 71, 96, 128
Meaning, 1, 7, 8, 12, 13, 15, 16, 19–23, 26, 27, 29, 31, 33, 35, 36, 39, 43, 47–50, 52, 53, 55, 59–73, 75, 78, 87–89, 91, 93–95, 98–103, 105, 108, 116, 119, 122–125, 127, 128, 131, 132, 135–139
Mediated, 14, 15, 57, 138
Medium, communication, 17, 18, 52

Methodological strategy, 3, 4, 54, 58, 112, 113
Miscarriage, 1, 5, 6, 30, 67, 69, 95, 99, 100, 126, 127
Moral, 16, 24, 29, 49, 70, 76, 87, 97, 108, 124
Mutual, mutuality, 2, 10, 13, 14, 31, 36, 41, 44, 45, 50, 52, 53, 77, 78

N

Narrative, 14, 15, 20, 47, 49, 52, 69, 73, 109, 130
Navigate, 35, 50, 53, 63, 66, 67, 72–74, 77, 109, 122, 131, 132, 137, 138
Negotiate, 49, 50, 63, 64, 66, 67, 73–75, 97, 122, 127, 131, 132, 137, 138
Network, inferential, identity, 59, 125
Neutrality, 23, 79, 85, 87–89, 93, 99, 136, 138
Niebuhr, H.R., 3, 48, 49, 53, 66, 68, 136
Nondirective, v, 3, 7, 8, 13, 23, 24, 27, 34, 40, 48, 55, 70, 78–103, 125, 126, 135, 136
Noninferential, 68
Nonjudgmental, 24, 86, 87, 92
Nonverbal, 26, 30, 36, 37, 39, 44, 93
Normativity, 3, 4, 7, 8, 11, 12, 17, 43, 49–52, 54, 56–58, 61, 63, 66, 67, 74, 80, 81, 84, 88, 91–93, 98, 101, 106, 114, 124, 130, 136, 138

O

Objective, 10, 19, 22, 23, 26, 27, 29, 30, 32, 52, 63, 67, 69, 73, 75, 81, 85, 88, 98, 99, 126, 132, 138
Obligation(s), 55, 57, 69, 70, 75, 76, 97, 98, 101, 111
Otherness, 78

P

Parens, E., 7, 49, 95
Pedagogy, 12, 27, 85, 87–89, 93, 136
Perception, 10, 28–30, 33, 37, 41, 70, 71, 115, 116, 119, 135
Performance, 5, 11, 41, 56, 57, 60, 62, 73, 85, 114, 119, 137
Persuasive, coercion, communication, 91, 93, 101
Peters, J.D., 1, 3, 4, 7, 13–18, 32, 40, 45, 47, 51–55, 72, 78, 118, 120, 132, 135, 136
Pluralistic society, 50, 97, 108
Plurality, 22, 28, 49, 50, 63, 97, 108, 109
Political norm, 97

Pragmatic, theory of communication, v, 3, 7, 51, 54–66
Prayer, 108, 110, 113, 119, 130, 139
Preference, patient, 14, 35, 69, 72, 97, 123
Pregnancy, 1, 2, 5–7, 30, 44, 49–51, 53, 60, 61, 69, 71, 79, 83, 90, 95–97, 99
Prenatal diagnosis, 1, 6, 7, 49, 125
Pro-attitudes, 69, 70
Probability, 6, 12, 23, 27, 30, 32, 50, 53, 74, 75, 84, 96, 99, 125, 129, 135
Process, dialogical, 45, 48, 50, 64, 65, 67
Psychosocial, 11, 24, 25, 28, 37, 41, 55, 70, 73, 75, 80, 89, 103, 113, 116–119, 123, 124, 127, 128
Psychotherapeutic model of genetic counseling, 3, 9–13, 36, 37, 39–43, 45, 47, 51, 68, 70, 71, 77, 80, 87, 89, 90, 93, 99–101, 124, 126–130, 132, 136

R

Randomness, 42, 111, 117
Rational, 10, 29, 30, 43, 47, 67, 68, 73, 82, 86, 88, 89, 100
Reasoning, practical, 29, 50, 73, 75–77, 94, 97
Recognition, 3, 13, 16, 17, 19, 21, 23, 27, 28, 33, 35, 36, 38, 39, 42, 44, 45, 48, 50, 52, 53, 55, 60, 71, 72, 77, 90, 103, 105, 106, 110, 127, 136
Reed, S., 23, 79, 80, 82, 85
Religion, 8, 55, 77, 99, 102–110, 112–114, 117–120, 123, 125–128, 131, 132, 137–139
Replication of self, 18, 51
Respect, 22, 26, 29, 37, 38, 61, 85, 86, 111, 117, 123, 130, 132, 136
Responsibility, model, shared, 49–51, 53, 54, 64, 66, 68, 77
Responsive, 13, 36, 43, 68, 71, 95, 109, 124
Risks, 1, 2, 5, 6, 12, 43, 44, 50, 51, 53, 55, 58, 60, 62, 64, 65, 67, 69, 71, 74, 76, 83, 85, 90, 91, 94–96, 98–100, 121, 123–126, 129, 131

S

Sanction, 56–58, 61
Schenck, D., 71, 125
Scorekeeping, deontic, 55–57, 63, 66, 77, 136, 137
Semantic, 16, 20, 22, 26, 27, 29, 31, 35, 39, 56, 59, 63, 64, 66, 68, 74, 89, 99–101, 127

Shared, decision making, responsibility, 13, 50
Sheridan, D., 106–109, 125, 136, 138
Sorenson, J., 3, 7, 23, 82, 85–88
Spatiotemporal, 2, 14, 17, 18, 25, 65
Spirit, Hegelian picture, 51
Spiritual assessment, 8, 78, 103–133, 135–137, 139–140
Spiritualist tradition, 4, 13–15, 17, 18, 43, 45, 47, 51, 52, 54, 77, 135
Spirituality, plenum, axial, 8, 103, 104
Sterilization, 81, 82
Stout, J., 75, 76, 125
Substitution, of words, 65
Symbols, 15, 19, 33, 35–37, 107

T

Teaching model of genetic counseling, 13, 21, 23–27
Technical vision of communication, v, 19–32, 47, 88, 99, 127
Termination of pregnancy, 90, 95, 97
Therapeutic vision of communication, v, 13, 14, 32–45

Tradition, v, 3, 4, 13–19, 27, 31, 41, 76, 83, 87, 90, 106, 107, 119, 135
Transcendent, 105, 107, 109, 125
Transform, human predicament, 107, 109, 125, 138
Transmission, 13, 15, 17, 18, 20, 21, 23, 26–28, 30, 45, 78, 100

U

Uncertainty, 7, 75, 83, 106, 108, 109, 116, 123, 125, 127
Unconditional, positive regard, obligation, 34, 36, 37, 70
Unmediated contact, 18

W

Weil, J., 3, 36–39, 41, 42, 44, 68, 80, 84, 89, 90, 93, 128–130, 132
White, M., 3, 7, 48, 49, 64, 66, 68, 70, 92, 97, 131, 136
Wittgenstein, L., 56